Method Development and Applications for Reduced-Risk Products in Separation Science

Method Development and Applications for Reduced-Risk Products in Separation Science

Editor

Fadi Aldeek

MDPI • Basel • Beijing • Wuhan • Barcelona • Belgrade • Manchester • Tokyo • Cluj • Tianjin

Editor
Fadi Aldeek
Altria Client Services LLC
USA

Editorial Office
MDPI
St. Alban-Anlage 66
4052 Basel, Switzerland

This is a reprint of articles from the Special Issue published online in the open access journal *Separations* (ISSN 2297-8739) (available at: https://www.mdpi.com/journal/separations/special_issues/method_application_risk).

For citation purposes, cite each article independently as indicated on the article page online and as indicated below:

LastName, A.A.; LastName, B.B.; LastName, C.C. Article Title. *Journal Name* **Year**, *Volume Number*, Page Range.

ISBN 978-3-0365-3637-8 (Hbk)
ISBN 978-3-0365-3638-5 (PDF)

Contents

About the Editor

Fadi Aldeek, Ph.D., serves as a Principal Scientist, Analytical Sciences, at Altria Client Services (ALCS) in Richmond, VA. In this role, he leads multiple analytical sciences research and development projects to expand analytical testing capabilities for novel oral nicotine-containing products and provides scientific insights to aid the development of reduced-risk products. Dr. Aldeek also provides technical oversights for other scientists and supports Altria's external engagement strategy. Dr. Aldeek's current research focuses on the development and validation of analytical methods to improve the analysis of constituents in reduced-risk products for regulatory reporting and for the support of product development. His research addresses all aspects of the analytical process, including sample preparation, extraction, separations, dissolution testing, detection, screening, quantification, identification, and data processing using a variety of state-of-the-art analytical technologies. Before joining the Altria family of companies in 2018, Dr. Aldeek was a Senior Scientist Group Leader at Eurofins Lancaster Laboratories (Richmond, Va.), leading an analytical chemistry team. Prior to this position, he served as a Senior Scientist at the Florida Department of Agriculture and Consumer Services where he established the Drug Residues Program to screen for antibiotics in various food matrices. Dr. Aldeek has also held positions at several academic institutions as a Research Fellow and Adjunct Professor. Dr. Aldeek is a co-author of more than 40 scientific peer-reviewed publications and 2 book chapters in his area of expertise. He is also co-author of over 60 presentations given at various national and international scientific meetings. Dr. Aldeek serves as associate and guest editor for the Journal of Nanomedicine Research, the Nanomaterials and Nanotechnology Journal, the Science Journal of Analytical Chemistry, and Separations.

Preface to "Method Development and Applications for Reduced-Risk Products in Separation Science"

Cigarette smoking is the most hazardous form of tobacco consumption due to the inherent risks of combusting tobacco and inhaling the smoke. Many in public health, including the FDA, agree that a continuum of risk exists among tobacco products, with cigarettes at the highest end and non-combustible tobacco products at the lower end of that continuum. Non-combustible products are lower on the continuum of risk because many of the harmful and potentially harmful constituents (HPHCs) found in cigarette smoke are either absent or present at very low levels. Switching to such products, therefore, may offer a harm reduction opportunity for adult smokers who cannot or will not quit smoking. In addition to the traditional smokeless tobacco products, non-combustible products also include innovative tobacco products such as oral tobacco-derived nicotine (OTDN) products, heated tobacco products (HTPs), and electronic cigarettes (also referred to as e-vapor products; EVPs).

Industry, academic, and government researchers are developing and validating analytical methods to extract, separate, identify, and quantitate a variety of analytes from these innovative tobacco products using a wide range of analytical techniques. These analytes include constituents such as nicotine, degradants and impurities, flavors, non-tobacco ingredients, HPHCs, and other currently unknown constituents.

In this Special Issue, we received nine contributions that covered the latest analytical methods that have been developed and applied for the chemical characterization or exposure assessment to tobacco product constituents of innovative non-combustible products. The developed methods included 1) characterizing the nicotine dissolution release profiles and determining nicotine degradants and HPHCs in OTDN pouches; 2) identifying HPHCs, targeted, and unknown compounds in EVPs; and 3) determining potential biomarkers at trace levels in urine and blood samples in these innovative products.

This Special Issue is representative of the importance of analytical sciences research in characterizing innovative non-combustible products for guiding product design, determining relative product performance, ensuring consistency during the manufacturing process, informing toxicological risk assessment, and enabling regulatory reporting.

The current advances in the development and applications of the analytical methods reported in this Special Issue can be used to inform the harm reduction potential of innovative non-combustible products for adult smokers.

I would like to take this opportunity to express my most profound appreciation to the MDPI Book staff, the editorial team of *Separations*, especially Mr. Ethan Xu, the assistant editor of this Special Issue, all of the talented authors, and the hardworking and professional reviewers.

Fadi Aldeek
Editor

 separations

Editorial

Method Development and Applications for Reduced-Risk Products

Fadi Aldeek * and Mohamadi A. Sarkar

Altria Client Services LLC, 601 East Jackson Street, Richmond, VA 23219, USA; mohamadi.a.sarkar@altria.com
* Correspondence: fadi.aldeek@altria.com

1. Introduction

Cigarette smoking remains the leading cause of preventable premature death and disease in the U.S. There is an overwhelming scientific consensus that cigarette smoking is addictive and causes lung cancer, heart disease, chronic obstructive pulmonary disease, and other serious diseases [1]. While there are thousands of constituents in cigarette smoke, ref. [2] certain representative classes of chemicals characterized as harmful and potentially harmful constituents (HPHCs) have been studied extensively and attributed to the harm caused by the inhaled smoke of combusted tobacco [3]. Many people in the public health sector have acknowledged that a continuum of risk exists among tobacco products, with conventional combustible cigarettes at the highest end of that spectrum, and non-combustible products on the lower end [4–6]. In recent years, there has been rapid growth in the availability of innovative, non-combustible products, including oral tobacco-derived nicotine (OTDN) products, heated tobacco products (HTPs), and electronic cigarettes (also referred to as e-vapor products; EVPs). Because they are non-combustible, such products contain far fewer combustion-related HPHCs [7–9]. As a result, substantial reduction in the biomarkers for exposure to HPHCs have been reported among adult smokers who completely switch to such products [10,11]. Such large reductions in exposure to HPHCs are accompanied with favorable changes in biomarkers indicative of smoking-related disease outcomes [12]. Consequently, there is a growing body of evidence suggesting that such products likely present a substantial reduction in disease risks [13], and many people in the public health sector recognize the potential of such non-combustible products for reducing harm [6,14,15]. Therefore, switching to non-combustible alternatives presents a significant opportunity to decrease the burden of disease associated with smoking combustible cigarettes, particularly among adult smokers who are unable or unwilling to quit.

There is a growing body of research dedicated to characterizing non-combustible products. Many researchers from industry, academia, and government are working to develop and validate analytical methods to extract, separate, identify, and quantitate a variety of analytes from innovative tobacco products using a wide range of analytical techniques. Understanding the basic properties of these products is important to better characterize innovative oral and inhalable tobacco products. The oral non-combustible categories include traditional smokeless tobacco and OTDN products. Traditional smokeless tobacco products contain tobacco leaves and exist in three different forms including chewing tobacco (loose leaf, plug, or twist); snuff (finely ground tobacco that can be dry, moist, or packaged in pouches (e.g., snus)); and dissolvable (finely ground tobacco pressed into shapes such as tablets, sticks, or strips) products [16]. OTDN products, on the other hand, are tobacco-leaf free and are available in various forms including nicotine pouches, lozenges, gums, and dissolvable products [17,18]. These products may contain a number of ingredients that include tobacco-derived nicotine, pH adjusters (e.g., sodium carbonates), filler materials (e.g., modified cellulose, microcrystalline cellulose), sweeteners, stabilizers, and flavorings.

Citation: Aldeek, F.; Sarkar, M.A. Method Development and Applications for Reduced-Risk Products. *Separations* **2022**, *9*, 78. https://doi.org/10.3390/separations9030078

Received: 16 March 2022
Accepted: 16 March 2022
Published: 18 March 2022

Publisher's Note: MDPI stays neutral with regard to jurisdictional claims in published maps and institutional affiliations.

Inhalable non-combustible products including EVPs and HTPs are compositionally different than cigarettes. Unlike traditional cigarettes, EVPs do not contain tobacco plant material or paper. They are mainly composed of a mixture of propylene glycol and glycerol in various ratios and flavors, and may or may not contain nicotine. In contrast, HTPs contain tobacco leaves but the tobacco is heated instead of burned, thereby lowering the temperature from >900 °C to ~500 °C. Due to the absence of tobacco leaves and paper in EVPs and the process of heating the tobacco in HTPs, many of the HPHCs in mainstream smoke are either not present or are present at significantly lower levels than smoking cigarettes [19,20].

The accurate determination and quantitation of constituents and chemicals in these products is needed for guiding product design, determining relative product performance, ensuring consistency during the manufacturing process, informing toxicological risk assessment, and regulatory reporting. This also allows for the characterization of inherent risks of innovative products, which helps determine whether the use of such products is potentially less harmful than smoking cigarettes. In this Special Issue, we discuss the latest analytical methods for chemical characterization of a variety of oral and inhalable non-combustible products.

2. Summary of Published Articles

This Special Issue includes research papers which address the latest analytical methods used for the identification and characterization of a variety of constituents and analytes in innovative oral and inhalable non-combustible tobacco products, using state-of-the-art techniques and instrumentations. The various contributions presented in this Special Issue are summarized based on the type of products evaluated and related methods reported.

Recently, nicotine pouches have emerged as a new category of innovative OTDN products. In this Special Issue, we received four contributions from different groups on methods that have been developed and validated to determine the nicotine release profiles, nicotine degradants, and HPHCs from a variety of nicotine pouch products. In these contributions, the authors have systematically used the developed methods to compare OTDN to traditional smokeless tobacco products. In the first manuscript, Aldeek et al. evaluated the nicotine release from 35 nicotine pouch products that are currently marketed in seven flavors with five different nicotine levels [21]. This is an important method to characterize the nicotine release from these pouches. The authors implemented a well-established dissolution method using the U.S. Pharmacopeia flow-through cell dissolution apparatus 4 (USP-4) that the same group previously developed for the evaluation of the nicotine release from traditional smokeless tobacco products [22]. The dissolution method was used for product-to-product comparison. The percent nicotine release profiles obtained from the 35 nicotine pouches under the same experimental conditions were found to be equivalent across all nicotine levels and flavors analyzed, indicating a similar rate of nicotine release from these oral nicotine pouch products. The authors further compared the percent nicotine release profiles from these nicotine pouches to a variety of other commercially available nicotine pouches and traditional pouched smokeless tobacco products. The authors state that the differences in percent nicotine release rates within the OTDN category could be associated with the inherent product characteristics (e.g., pouch paper and ingredients).

In the second manuscript, Knopp et al. developed a biorelevant dissolution method to study the nicotine release from OTDN nicotine pouches and portioned smokeless tobacco products (e.g., pouched snus) [23]. The in vitro release of nicotine was investigated in biorelevant volumes of artificial saliva using a custom-made dissolution apparatus. The apparatus consisted of a sinker that was prepared by 3D printing using polylactic acid material. The nicotine released was quantitated by a validated high-pressure liquid chromatography ultra-violet spectroscopy (HPLC-UV) method. The percent nicotine release profiles obtained from the OTDN and snus pouches were found to be distinct, indicating the ability of this method to discriminate between these two product categories. Additionally, the authors compared the in vitro dissolution to in vivo data from a previously conducted clinical study [24]. Data showed a strong in vitro/in vivo correlation, indicating that the method reported in this publication is not only sensitive enough to discriminate between nicotine pouch and snus products, but could also serve as a predictive tool for product development and/or a monograph for oral tobacco/nicotine product equivalence studies.

The stability of nicotine depends on the inherent components of the product (e.g., fillers, pH, stabilizers, other ingredients, and moisture content) as well as the external environment (e.g., exposure to light and high temperatures). Therefore, developing methods to assess the nicotine stability in these products by monitoring the nicotine degradation compounds and select impurities is very important. These methods are useful to monitor the stability of nicotine in these products and for quality control purposes (e.g., to evaluate the purity of nicotine added to the product). In the third manuscript, Avagyan et al. developed a selective, accurate, and repeatable liquid chromatography–tandem mass spectrometry (LC-MS/MS) method for the determination of seven nicotine-related degradants and impurities [25]. The seven nicotine degradants in this method were nicotine-N'-oxide, cotinine, nornicotine, anatabine, anabasine, ß-nicotyrin, and myosmine. Most of the analytes were detected in the nicotine pouch products; however, they were found to be at lower levels compared to traditional tobacco products.

In the fourth manuscript, Jablonski et al. used fully validated CORESTA recommended methods to determine 17 selected HPHCs (including tobacco-specific nitrosamines, carbonyls, benzo[a]pyrene, nitrite, and metals) from 21 nicotine pouch products [26]. The selected pouches were obtained from seven different commercially available brands at the maximum nicotine level and a variety of flavors. The authors assessed two types of pouch products described as "white powder-based pouches" and "plant-based" pouches. The white powder-based pouches were similar to those described above, whereas the plant-based pouches were made from non-tobacco plant materials with pharmaceutical grade nicotine added during the production process. HPHCs in the 21 nicotine pouches were compared to those found in four traditional smokeless tobacco products (two CORESTA reference products and two commercially available products). The authors reported that the HPHCs levels, most notably metals, in the plant-based pouches were higher than those observed in powder-based products. In some plant-based pouches, these levels were even higher than those seen in traditional pouch smokeless tobacco products. However, the overall HPHCs levels observed in these plant-based nicotine pouches were at or below those levels observed in traditional pouch smokeless tobacco products.

The presence of unique constituents in the aerosol of EVPs is an important consideration in overall risk assessment of such products and is of interest to regulators and public health researchers. EVPs include both the e-liquid (containing nicotine and other ingredients) and aerosolizing apparatus, whether sold as a unit or separately. Due to the unique parts and components of EVPs, the constituents are distinct and specific to the product type (e.g., pod-based, open system, etc.). Therefore, in addition to the HPHCs, unknown compounds in the aerosol need to be characterized. The majority of analytical work on EVPs has focused on targeting known chemicals of interest based on changes to the device, formulation, power, temperature, or sampling approaches [27]. In this Special Issue, we received three contributions highlighting the development of targeted and non-targeted analytical methods for the determination of HPHCs and unknowns in EVPs. In the first report, Jin et al. evaluated the traditional 2,4-dinitrophenylhdrazine (2,4-DNPH) derivatization and quantitation of formaldehyde in e-liquid and aerosol of EVPs [28]. Formaldehyde is an HPHC listed by the FDA as a carcinogen and a respiratory toxicant [3]. Previous reports stated that formaldehyde is often underreported in EVPs due to a possible reaction with propylene glycol and glycerin in the aerosol which causes the formation of hemiacetals [29]. The research presented in this study provided a thorough experimental design to clearly demonstrate that hemiacetals formed in the aerosol readily hydrolyze to free formaldehyde and consequently form formaldehyde hydrazone in the typical 2,4-DNPH acidic trapping solution for quantitation. This study showed that the commonly used 2,4-DNPH method is an appropriate method for the derivatization and accurate quantitation of formaldehyde in the aerosol generated by EVPs.

In the second manuscript, Chen et al. developed a comprehensive, targeted analysis using gas chromatography coupled to mass spectrometry (GC-MS) for the determination of 53 aerosol constituents from EVPs of currently marketed products [30]. The aerosol generation was conducted using non-intense and intense puffing regimens. Only 10 out of the targeted 53 analytes were quantifiable. The authors have compared their data to constituents collected from aerosols generated by both traditional cigarettes and a commercially available HTP that has been authorized for marketing in the U.S. The aerosol generated by the evaluated EVPs had detectable levels of ten targeted analytes including known degradants of propylene glycol and glycerin (e.g., acetaldehyde and formaldehyde) and nicotine-related compounds. The majority of tobacco-related HPHCs were not detectable in the aerosols. The levels of select HPHCs (other than nicotine) measured in the EVPs were found to be 96–99% lower than the same HPHCs reported in the cigarette smoke. However, the reduction levels of these select HPHCs in the EVPs ranged from 61% to 99% when compared to the levels found in HTP aerosol. The authors attributed the low levels of HPHCs in the EVPs' aerosols to the controlled temperature used in the device which is designed to reduce byproducts of combustion.

To address the potential gaps in understanding left by targeted analysis of EVPs, Crosswhite et al. developed and optimized liquid chromatography high resolution mass spectrometry (LC-HRMS) and GC-MS semi-quantitative methods to study unknown chemicals in generated aerosols [31]. These two methods were developed to account for the different physicochemical properties of possible chemical compounds including polarity, volatility, hydrophilicity, etc. The authors used differential analyses based on nine aerosol collection replicates of each studied EVP and each collection condition (intense and non-intense puffing regimens) to characterize compounds that differed from collection blanks. They relied on statistical tools to extract relevant information from a highly complex dataset. The authors reported all compounds at or above concentrations of 0.5 µg/g which were considered related to the sample. A total of 91 compounds were identified using these two methods in both non-intense and intense puffing regimens. This number was strikingly low when compared to the number of compounds (>5000) found in cigarette smoke [32]. Of the detected compounds, 47% were confirmed using reference standards. The authors showed that the studied aerosols from EVPs were approximately 50-fold less complex when compared to cigarette smoke.

We have also received two articles describing the development of LC-MS/MS methods for the identification of biomarkers of exposure specific to EVPs and other non-combustible products. Burkhardt et al. developed an LC-MS/MS method for measuring human exposure to 1,2-propylene glycol and glycerol, the main e-liquid constituents in EVPs [33]. These constituents were analyzed in plasma and urine samples from a clinical study comparing five nicotine product user groups (users of combustible cigarettes, EVPs, HTPs, oral tobacco products, and nicotine replacement therapy (NRT) products) and a control group of non-users. The results demonstrated elevated propylene glycol levels in urine and plasma in EVPs users compared to users of other products. The data showed a correlation between the propylene glycol and nicotine equivalents in the plasma and urine of EVP users. The nicotine equivalents were calculated by measuring the levels of nicotine and ten nicotine metabolites using a method developed by Piller et al. [34]. The authors also reported a dose–response relationship between urinary and plasma propylene glycol and intensity of vaping. The authors proposed that propylene glycol can be used as a potential biomarker to monitor compliance to EVP use when assessing switching behavior among smokers.

The same group, in a second article by Rogner et al., developed and validated another highly sensitive LC-MS/MS method for the determination of 3-hydroxybenzo[a]pyrene (3-OH-BaP), a metabolite of benzo[a]pyrene (BaP), in urine samples from smokers and non-combustible products users [35]. BaP is listed by FDA as an HPHC and classified by IARC as a human carcinogen which is formed during the incomplete combustion of tobacco [3]. The method was validated with a very low limit of quantitation (50 pg/L) to account for trace levels of 3-OH-BaP in urine samples. The detected levels of 3-OH-BaP in urine samples were found to be significantly higher in cigarette smokers compared to non-combustible product users. The data presented by the authors showed the suitability of 3-OH-BaP as a biomarker for BaP and could be applied in clinical studies evaluating innovative non-combustible tobacco products.

3. Conclusions

The nine articles published in this Special Issue covered the latest analytical methods developed and applied for the chemical characterization or exposure assessment to tobacco product constituents of innovative non-combustible products (i.e., EVPs, HTPs, and OTDN products). The developed methods included (1) characterizing the nicotine dissolution release profiles and determining nicotine degradants and HPHCs in OTDN pouches; (2) identifying HPHCs, targeted, and unknown compounds in EVPs; and (3) determining potential biomarkers at trace levels in urine and blood samples in a variety of EVPs, HTPs, and OTDN products. The contributors to this Special Issue systematically compared the amount and release characteristics of select HPHCs, degradants, and unknown compounds found in innovative non-combustible products to combustible cigarettes or traditional smokeless tobacco products. This Special Issue is representative of the importance of analytical sciences research in characterizing innovative non-combustible products for guiding product design, determining relative product performance, ensuring consistency during the manufacturing process, informing toxicological risk assessment, and enabling regulatory reporting. The current advances in the development and applications of the analytical methods reported in this Special Issue can be used to inform the harm reduction potential of innovative non-combustible products for adult smokers.

Author Contributions: Conceptualization, F.A.; writing—review and editing, F.A. and M.A.S. All authors have read and agreed to the published version of the manuscript.

Funding: This research received no external funding.

Acknowledgments: Fadi Aldeek thanks all the authors for their excellent contributions. The efforts of the reviewers are acknowledged as contributing greatly to the quality of this Special Issue.

Conflicts of Interest: The authors declare no conflict of interest.

References

1. US Department of Health and Human Services. *The Health Consequences of Smoking—50 Years of Progress: A Report of the Surgeon General*; Centers for Disease Control and Prevention (US): Atlanta, GA, USA, 2014.
2. Rodgman, A.; Perfetti, T.A. *The Chemical Components of Tobacco and Tobacco Smoke*; CRC Press: Boca Raton, FL, USA, 2013.
3. FDA. *Harmful and Potentially Harmful Constituents in Tobacco Products and Tobacco Smoke: Established List*; Office of the Federal Register: Washington, DC, USA, 2012.
4. Gottlieb, S.; Zeller, M. A Nicotine-Focused Framework for Public Health. *N. Engl. J. Med.* **2017**, *377*, 1111–1114. [CrossRef] [PubMed]
5. Hatsukami, D.K.; Joseph, A.M.; Lesage, M.; Jensen, J.; Murphy, S.E.; Pentel, P.R.; Kotlyar, M.; Borgida, E.; Le, C.; Hecht, S.S. Developing the science base for reducing tobacco harm. *Nicotine Tob. Res.* **2007**, *9*, 537–553. [CrossRef] [PubMed]
6. Zeller, M.; Hatsukami, D. The Strategic Dialogue on Tobacco Harm Reduction: A vision and blueprint for action in the US. *Tob. Control* **2009**, *18*, 324–332. [CrossRef] [PubMed]
7. Wagner, K.A.; Flora, J.W.; Melvin, M.S.; Avery, K.C.; Ballentine, R.M.; Brown, A.P.; McKinney, W.J. An evaluation of electronic cigarette formulations and aerosols for harmful and potentially harmful constituents (HPHCs) typically derived from combustion. *Regul. Toxicol. Pharmacol.* **2018**, *95*, 153–160. [CrossRef]
8. Jaccard, G.; Djoko, D.T.; Moennikes, O.; Jeannet, C.; Kondylis, A.; Belushkin, M. Comparative assessment of HPHC yields in the Tobacco Heating System THS2.2 and commercial cigarettes. *Regul. Toxicol. Pharmacol.* **2017**, *90*, 1–8. [CrossRef]
9. Danielson, T.M.; Brown, A.P.; Jin, X.; Wilkinson, C.T.; Pithawalla, Y.; McKinney, W.J. Evaluation of Novel, Oral Tobacco-Derived Nicotine Products for HPHCs. In Proceedings of the 72nd Tobacco Science Research Conference, Memphis, TN, USA, 16–19 September 2018.

10. Haziza, C.; De La Bourdonnaye, G.; Donelli, A.; Poux, V.; Skiada, D.; Weitkunat, R.; Baker, G.; Picavet, P.; Lüdicke, F. Reduction in Exposure to Selected Harmful and Potentially Harmful Constituents Approaching Those Observed Upon Smoking Abstinence in Smokers Switching to the Menthol Tobacco Heating System 2.2 for 3 Months (Part 1). *Nicotine Tob. Res.* **2019**, *22*, 539–548. [CrossRef]
11. Sarkar, M.; Kapur, S.; Frost-Pineda, K.; Feng, S.; Wang, J.; Liang, Q.; Roethig, H. Evaluation of biomarkers of exposure to selected cigarette smoke constituents in adult smokers switched to carbon-filtered cigarettes in short-term and long-term clinical studies. *Nicotine Tob. Res.* **2008**, *10*, 1761–1772. [CrossRef]
12. Lüdicke, F.; Ansari, S.M.; Lama, N.; Blanc, N.; Bosilkovska, M.; Donelli, A.; Picavet, P.; Baker, G.; Haziza, C.; Peitsch, M.; et al. Effects of Switching to a Heat-Not-Burn Tobacco Product on Biologically Relevant Biomarkers to Assess a Candidate Modified Risk Tobacco Product: A Randomized Trial. *Cancer Epidemiol. Prev. Biomark.* **2019**, *28*, 1934–1943. [CrossRef]
13. Polosa, R.; Farsalinos, K.; Prisco, D. Health impact of electronic cigarettes and heated tobacco systems. *Intern. Emerg. Med.* **2019**, *14*, 817–820. [CrossRef] [PubMed]
14. Hajek, P. Electronic cigarettes have a potential for huge public health benefit. *BMC Med.* **2014**, *12*, 225. [CrossRef]
15. Balfour, D.J.K.; Benowitz, N.L.; Colby, S.M.; Hatsukami, D.K.; Lando, H.A.; Leischow, S.J.; Lerman, C.; Mermelstein, R.J.; Niaura, R.; Perkins, K.A.; et al. Balancing Consideration of the Risks and Benefits of E-Cigarettes. *Am. J. Public Health* **2021**, *111*, 1661–1672. [CrossRef] [PubMed]
16. Food and Drug Administration, Smokeless Tobacco Products, including Dip, Snuff, Snus, and Chewing Tobacco. Available online: https://www.fda.gov/tobacco-products/products-ingredients-components/smokeless-tobacco-products-including-dip-snuff-snus-and-chewing-tobacco (accessed on 14 March 2022).
17. Choi, J.H.; Dresler, C.M.; Norton, M.R.; Strahs, K.R. Pharmacokinetics of a nicotine polacrilex lozenge. *Nicotine Tob. Res.* **2003**, *5*, 635–644. [CrossRef] [PubMed]
18. Robichaud, M.O.; Seidenberg, A.B.; Byron, M.J. Tobacco companies introduce 'tobacco-free' nicotine pouches. *Tob. Control* **2019**, *29*, e145–e146. [CrossRef]
19. Tayyarah, R.; Long, G.A. Comparison of select analytes in aerosol from e-cigarettes with smoke from conventional cigarettes and with ambient air. *Regul. Toxicol. Pharmacol.* **2014**, *70*, 704–710. [CrossRef] [PubMed]
20. Gillman, I.; Kistler, K.; Stewart, E.; Paolantonio, A. Effect of variable power levels on the yield of total aerosol mass and formation of aldehydes in e-cigarette aerosols. *Regul. Toxicol. Pharmacol.* **2016**, *75*, 58–65. [CrossRef]
21. Aldeek, F.; McCutcheon, N.; Smith, C.; Miller, J.H.; Danielson, T.L. Dissolution Testing of Nicotine Release from OTDN Pouches: Product Characterization and Product-to-Product Comparison. *Separations* **2021**, *8*, 7. [CrossRef]
22. Miller, J.H.; Danielson, T.; Pithawalla, Y.B.; Brown, A.P.; Wilkinson, C.; Wagner, K.; Aldeek, F. Method development and validation of dissolution testing for nicotine release from smokeless tobacco products using flow-through cell apparatus and UPLC-PDA. *J. Chromatogr. B Analyt. Technol. Biomed. Life Sci.* **2020**, *1141*, 122012. [CrossRef]
23. Knopp, M.M.; Kiil-Nielsen, N.K.; Masser, A.E.; Staaf, M. Introducing a Novel Biorelevant In Vitro Dissolution Method for the Assessment of Nicotine Release from Oral Tobacco-Derived Nicotine (OTDN) and Snus Products. *Separations* **2022**, *9*, 52. [CrossRef]
24. A Study Investigating the Extraction of Nicotine and Flavors from Tobacco Free Nicotine Pods Compared to Tobacco Based Swedish Snus. Available online: https://www.isrctn.com/ISRCTN44913332?q=&filters=conditionCategory:Not%20Applicable&page=1&pageSize=10 (accessed on 14 March 2022).
25. Avagyan, R.; Spasova, M.; Lindholm, J. Determination of Nicotine-Related Impurities in Nicotine Pouches and Tobacco-Containing Products by Liquid Chromatography–Tandem Mass Spectrometry. *Separations* **2021**, *8*, 77. [CrossRef]
26. Jablonski, J.J.; Cheetham, A.G.; Martin, A.M. Market Survey of Modern Oral Nicotine Products: Determination of Select HPHCs and Comparison to Traditional Smokeless Tobacco Products. *Separations* **2022**, *9*, 65. [CrossRef]
27. Ward, A.M.; Yaman, R.; Ebbert, J.O. Electronic nicotine delivery system design and aerosol toxicants: A systematic review. *PLoS ONE* **2020**, *15*, e0234189. [CrossRef]
28. Jin, X.C.; Ballentine, R.M.; Gardner, W.P.; Melvin, M.S.; Pithawalla, Y.B.; Wagner, K.A.; Avery, K.C.; Sharifi, M. Determination of Formaldehyde Yields in E-Cigarette Aerosols: An Evaluation of the Efficiency of the DNPH Derivatization Method. *Separations* **2021**, *8*, 151. [CrossRef]
29. Salamanca, J.C.; Munhenzva, I.; Escobedo, J.O.; Jensen, R.P.; Shaw, A.; Campbell, R.; Luo, W.; Peyton, D.H.; Strongin, R.M. Formaldehyde Hemiacetal Sampling, Recovery, and Quantification from Electronic Cigarette Aerosols. *Sci. Rep.* **2017**, *7*, 11044. [CrossRef] [PubMed]
30. Chen, X.; Bailey, P.C.; Yang, C.; Hiraki, B.; Oldham, M.J.; Gillman, I.G. Targeted Characterization of the Chemical Composition of JUUL Systems Aerosol and Comparison with 3R4F Reference Cigarettes and IQOS Heat Sticks. *Separations* **2021**, *8*, 168. [CrossRef]
31. Crosswhite, M.R.; Bailey, P.C.; Jeong, L.N.; Lioubomirov, A.; Yang, C.; Ozvald, A.; Jameson, J.B.; Gillman, I.G. Non-Targeted Chemical Characterization of JUUL Virginia Tobacco Flavored Aerosols Using Liquid and Gas Chromatography. *Separations* **2021**, *8*, 130. [CrossRef]
32. Green, C.R.; Rodgman, A. The Tobacco Chemists' Research Conference: A half century forum for advances in analytical methodology of tobacco and its products. *Recent Adv. Tob. Sci.* **1996**, *22*, 131–304.
33. Burkhardt, T.; Pluym, N.; Scherer, G.; Scherer, M. 1,2-Propylene Glycol: A Biomarker of Exposure Specific to e-Cigarette Consumption. *Separations* **2021**, *8*, 180. [CrossRef]

34. Piller, M.; Gilch, G.; Scherer, G.; Scherer, M. Simple, fast and sensitive LC–MS/MS analysis for the simultaneous quantification of nicotine and 10 of its major metabolites. *J. Chromatogr. B* **2014**, *951–952*, 7–15. [CrossRef]
35. Rögner, N.; Hagedorn, H.-W.; Scherer, G.; Scherer, M.; Pluym, N. A Sensitive LC–MS/MS Method for the Quantification of 3-Hydroxybenzo[*a*]pyrene in Urine-Exposure Assessment in Smokers and Users of Potentially Reduced-Risk Products. *Separations* **2021**, *8*, 171. [CrossRef]

 *separations*

MDPI

Article

Dissolution Testing of Nicotine Release from OTDN Pouches: Product Characterization and Product-to-Product Comparison

Fadi Aldeek *, Nicholas McCutcheon, Cameron Smith, John H. Miller and Timothy L. Danielson

Altria Client Services LLC, 601 East Jackson Street, Richmond, VA 23219, USA;
Nicholas.McCutcheon@altria.com (N.M.); Cameron.R.Smith@altria.com (C.S.);
John.H.Miller@altria.com (J.H.M.); Timothy.L.Danielson@altria.com (T.L.D.)
* Correspondence: fadi.aldeek@altria.com; Tel.: +1-804-335-3119

Abstract: In recent years, oral tobacco-derived nicotine (OTDN) pouches have emerged as a new oral tobacco product category. They are available in a variety of flavors and do not contain cut or ground tobacco leaf. The on!® nicotine pouches fall within this category of OTDN products and are currently marketed in seven (7) flavors with five (5) different nicotine levels. Evaluation of the nicotine release from these products is valuable for product assessment and product-to-product comparisons. In this work, we characterized the in vitro release profiles of nicotine from the 35 varieties of on!® nicotine pouches using a fit-for-purpose dissolution method, employing the U.S. Pharmacopeia flow-through cell dissolution apparatus 4 (USP-4). The nicotine release profiles were compared using the FDA's Guidance for Industry: *Dissolution Testing of Immediate Release Solid Oral Dosage Forms.* The cumulative release profiles of nicotine show a dose dependent response for all nicotine levels. The on!® nicotine pouches exhibit equivalent percent nicotine release rates for each flavor variant across all nicotine levels. Furthermore, the nicotine release profiles from on!® nicotine pouches were compared to a variety of other commercially available OTDN pouches and traditional pouched smokeless tobacco products. The percent nicotine release rates were found to be dependent on the product characteristics, showing similarities and differences in the nicotine release profiles between the on!® nicotine pouches and other compared products.

Keywords: on!® nicotine pouches; nicotine; dissolution; release profile; validation; product assessment; smokeless tobacco product

Citation: Aldeek, F.; McCutcheon, N.; Smith, C.; Miller, J.H.; Danielson, T.L. Dissolution Testing of Nicotine Release from OTDN Pouches: Product Characterization and Product-to-Product Comparison. *Separations* **2021**, *8*, 7. https://doi.org/10.3390/separations8010007

Received: 15 October 2020
Accepted: 2 January 2021
Published: 7 January 2021

Publisher's Note: MDPI stays neutral with regard to jurisdictional clai-ms in published maps and institutio-nal affiliations.

1. Introduction

Over recent years, oral tobacco products have provided alternatives to smoking cigarettes [1,2]. The use of oral tobacco products is considered by many to have potentially reduced risks of harm compared to smoking cigarettes [3–5]. Oral tobacco products exist in two major categories: traditional smokeless tobacco and modern oral nicotine products. Typically, traditional smokeless tobacco products come in three different types, including chewing tobacco (loose leaf, plug, or twist), snuff (finely ground tobacco that can be dry, moist, or packaged in pouches (e.g., snus)), and dissolvable (finely ground tobacco pressed into shapes such as tablets and sticks) products [6]. While traditional smokeless tobacco products contain tobacco leaves, modern oral nicotine products are tobacco leaf-free that contain tobacco-derived nicotine and food grade ingredients [7]. In the last decade, modern oral tobacco-derived nicotine (OTDN) products have been commercialized in various solid forms, including lozenges, gums, and dissolving tablets [8–10]. More recently, nicotine pouches have emerged as a new category of OTDN products. These products are pre-portioned pouches similar to snus but replace the tobacco leaf with non-tobacco filler and tobacco-derived nicotine.

The scientific evidence regarding the long-term health effects of OTDN pouches has not yet been established; however, the vast body of literature on other oral tobacco products, such as moist smokeless tobacco products, suggests that nicotine pouches will pose significantly lower risks than cigarettes [11]. Based on our review of statements from authoritative bodies regarding the long-term health effects of nicotine and available scientific literature on nicotine replacement therapy (NRT) as well as moist smokeless tobacco products, we believe that OTDN pouch products are not risk free and can lead to dependence. Nicotine, while not benign, has substantially lower health risks compared to smoking cigarettes [12,13].

Tobacco product manufacturers are currently selling OTDN pouch products under different brand names such as ZYN®, Velo, and on!® [7]. These products come in a variety of flavors and different nicotine contents per pouch. The on!® nicotine pouch products, for example, are currently marketed with seven flavor variants (e.g., Citrus, Wintergreen, Mint, Coffee, Berry, Cinnamon, and Original) and 5 different nicotine levels (1.5, 2, 3.5, 4, and 8 mg per pouch), providing an overall portfolio of 35 combinations of flavor variants and nicotine levels. These products are consumed by placing the pouch between the gum and upper lip, allowing for the dissolution of nicotine to occur in the saliva before being absorbed in the oral cavity and entering the bloodstream [14].

The market for oral nicotine pouches has been increasing in recent years as adult tobacco consumers are looking for alternatives to more traditional tobacco products, such as cigarettes [4]. Therefore, research evaluating the release of nicotine from these pouches is needed for product characterization and product-to-product comparisons.

Dissolution testing is commonly used by the pharmaceutical industry to assess product quality, demonstrate equivalency in constituent release, guide formulation design, and develop in vivo/in vitro correlation (IVIVC) [15–19]. Dissolution testing measures in vitro drug release as a function of time, which may reflect the reproducibility of the manufacturing process and, in some cases, relates to the active ingredient's in vivo release [20–23]. Despite the numerous well-established and standardized methods described in the pharmacopoeias, only a few dissolution methods have been developed for the comparison of OTDN, using a variety of dissolution apparatus and analytical methods [24–27]. Recently, we developed and validated a fit-for-purpose method for the dissolution testing of nicotine from a variety of traditional smokeless tobacco products using a USP-4 flow-through cell apparatus. This method quantitatively determines the nicotine release into artificial saliva from a variety of smokeless tobacco products selectively and precisely. This discriminatory dissolution methodology was successfully applied to study the dissolution release profiles from a variety of traditional reference and commercial smokeless tobacco products. We demonstrated the ability of this method to be used as an important tool for tobacco product assessment and product-to-product comparisons, and also that the nicotine release profile is dependent on the form and cut of the studied traditional smokeless tobacco products [28].

In this study, we built on our initial findings and expanded the scope of our validated method to include oral nicotine pouch products, on!®. We characterized the dissolution release of nicotine from 35 on!® nicotine pouch products across the seven flavors and five nicotine levels by comparing the cumulative and percent of total nicotine release profiles. We further calculated the difference factor (f_1) and similarity factor (f_2) using a methodology referenced in the Guidance for Industry from FDA's Center for Drug Evaluation and Research (CDER) [29,30]. Furthermore, the nicotine release profiles from on!® nicotine pouches were compared to a variety of OTDN pouches and traditional smokeless tobacco products to better understand the nicotine release rates within and across product categories.

2. Materials and Methods

The dissolution testing was carried out using a USP-4 flow-through cell apparatus (SOTAX, Westborough, MA, USA) following our previous methodology [28]. The determination of nicotine was performed using Acquity I-Class Ultra Performance Liquid Chromatography coupled to a Photodiode Array detector (UPLC-PDA) (Waters, Milford, MA, USA). The UPLC was fitted with a BEH C18 analytical column (2.1 × 100 mm, 1.7 µm) and a BEH C18 VanGuard pre-column (2.1 × 5 mm, 1.7 µm) (Waters, Milford, MA,

USA) [28]. The artificial saliva was prepared according to the method described in the German Institute for Standardization (DIN) recipe listed in the German standard DIN V Test Method 53160-1 2002-10 [31]. The USP-4 fractions collection and UPLC solutions and standards preparation were performed following our previously published report [28].

2.1. Test Products

The 35 on!® nicotine pouch products, currently marketed with seven flavor variants (Citrus, Wintergreen, Mint, Coffee, Berry, Cinnamon, and Original) and 5 different nicotine levels (1.5, 2, 3.5, 4, and 8 mg per pouch) were provided by the manufacturer. Similarly, the Skoal® Bandits and Skoal® pouches (Wintergreen flavored traditional pouched smokeless tobacco products) were provided by the manufacturer. The ZYN® nicotine pouch products used in this study, with different flavor variants (Coffee, Wintergreen, and Cool Mint) and nicotine levels (3 and 6 mg per pouch), were purchased from retail stores.

2.2. Dissolution Fractions Collection

The USP-4 apparatus used in this study consisted of an array of seven flow-through cells, a cell holder including a water bath, a reservoir and pump for artificial saliva, and a fractions collection rack. The pump delivered a constant flow of artificial saliva (4 mL/min) through the flow-through cells. The flow-through cells were mounted vertically with a filter system that prevents the pouches from exiting the cell. The cells were immersed in the water bath, and the temperature was maintained at 37 ± 0.5 °C. A 5 mm ruby bead check valve was placed in the bottom of each sample cell, and approximately 6.6 g of 1 mm glass beads was added to the conical portion of the cell to ensure a laminar flow. Pouched products were weighed, and a single pouch of on!® (~0.265 g), ZYN® (~0.393 g), or traditional pouched smokeless tobacco product (~0.72–1.55 g), was added directly into each vessel. The cell was then filled with approximately 6.6 g of 3 mm glass beads to maintain the pouch position in the center of the flow-through cell. The dissolution testing was conducted according to the guidance issued by the FDA using 12 replicates of one product and taking a dissolution profile at a maximum of 15-min intervals [29,30]. Each replicate was dissolved into 9 fractions. The collection time was 4 min for fractions 1–5, resulting in a final collection volume of 16 mL for each fraction, and 10 min for fractions 6–9, resulting in a final collection volume of 40 mL in each fraction for a total dissolution time of 60 min.

2.3. Quantitative Analysis of Nicotine

Upon collection of all 9 fractions from each sample replicate, 0.1 mL of each dissolution fraction was added to an autosampler vial, followed by the addition of 0.1 mL of ethyl benzoate as an internal standard (1 mg/mL) and 0.8 mL of artificial saliva. The nicotine concentration in µg/mL was quantitated in all fractions collected from the 12 replicates following the analytical UPLC-PDA method described previously [28]. The concentration of the nicotine based on sample pouch (nicotine amount (mg) released), was determined using the calculated concentration of nicotine (µg/mL), weight of the sample analyzed, and volume of the dissolution fraction.

2.4. Cumulative and Percent of Total Release Profiles

The cumulative concentrations of nicotine (nicotine amount (mg) released) from each tested product were calculated by summing the averaged nicotine released for each fraction time point from all 12 replicates. The sum of the averaged cumulative nicotine amount corresponds to the total amount of nicotine released up to each time point. The percentage relative to the total nicotine released at each time point (rate) was then calculated and plotted to provide the total release profile. The relative percentage to the total nicotine released was calculated by dividing the amount of nicotine released up to each time point for each fraction by the cumulative amount released in 60 min.

2.5. F1 and F2 Calculations

The difference factor (f_1) and similarity factor (f_2) were calculated by adopting a methodology referenced in the Guidance for Industry from FDA's Center for Drug Evaluation and Research (CDER) [29,30]. These two factors can be calculated mathematically by the following equations [32,33]:

$$f_1 = \left\{ \left[\sum_{t=1}^{n} |R - T| \right] / \left[\sum_{t=1}^{n} R \right] \right\} \times 100 \tag{1}$$

$$f_2 = 50 \cdot \log \left[\frac{100}{\sqrt{1 + \frac{\sum_{t=1}^{t=n} [Rt - Tt]^2}{n}}} \right] \tag{2}$$

R_t and T_t are the cumulative percentage dissolved at each of the selected n time points of the two products. The factor f_1 is proportional to the average difference between the two profiles, whereas factor f_2 is inversely proportional to the average squared difference between the two profiles, with emphasis on the larger difference among all the time points. Following the FDA's guidance document, at least 12 replicates should be used for each profile determination. The dissolution measurements of the two products should also be made under identical test conditions. For curves (kinetic release profiles) to be considered equivalent, f_1 values should be close to 0 and f_2 values should be close to 100. Generally, f_1 values up to 15 (0–15) and f_2 values of 50 or greater (50–100) demonstrate equivalence of the two curves, reflecting a similar performance of the two products.

3. Results and Discussion

Previously, we developed and validated a dissolution method to quantitatively evaluate the rate of nicotine release from traditional smokeless tobacco products using USP-4 flow-through cell dissolution apparatus and UPLC-PDA. We based our approach on consensus methodology already existing in the field of pharmaceutical products, including the choice of apparatus, dissolution medium, and the analytical method used for the nicotine quantitation [26,34,35]. We described approaches for product-to-product comparisons between various nicotine-containing traditional loose and pouched traditional smokeless tobacco products [28]. Here, we expanded this methodology to measure the rate of nicotine release for the on!® nicotine pouches portfolio, consisting of 35 products (7 flavors at 5 different nicotine levels).

3.1. Method Validation

Our USP-4 flow-through cell/UPLC-PDA method was initially validated to study the dissolution release of nicotine from loose and pouched traditional smokeless tobacco products. To study the nicotine release profile from on!® nicotine pouch products, we conducted a supplemental validation to expand the scope of our original method. The supplemental validated elements of the method were accuracy, precision, specificity, and fraction stability. Accuracy of the analytical method was measured by calculating the recovery from two fortification levels in pooled fractions collected from 1.5 mg and 8 mg on!® nicotine pouch products of all flavor variants. Dissolution fractions from the beginning (fractions 1–5) and end (fractions 6–9) of the collection were combined into two pools: pool #1 (fractions 1–5) and pool #2 (fractions 6–9). The fortification levels were 50 and 200 µg/mL for pool #1 and 10 and 50 µg/mL for pool #2. Three replicates of each fortified sample were analyzed to determine accuracy. To determine the % recovery, the measured nicotine value from the unfortified samples was subtracted from each of the fortified samples. The corresponding results were divided by the fortified amounts to determine % recovery. All fortification levels and matrix types had calculated nicotine recovery values between 85 and 107%. Intra-day precision was determined by analyzing 3 replicates each of on!® Mint 1.5 mg and 8 mg pouches within a single day and was found to be <3% Relative Standard Deviation (RSD). Intermediate precision was measured by analyzing 3 replicates each of the same product over the course of three days ($n = 9$) and was found to be <4% RSD. The specificity of the method was validated by examining the chromatograms in all fractions and artificial saliva (used as a blank). The chromatograms were free of matrix interference, showing the ability of the method to quantitate nicotine in this sample matrix. Finally, the stability of the dissolution fractions was assessed over a period of 14 days in amber glass bottles with a screw cap at 0–4 °C. An initial analysis was made for time zero (day 1) and compared to the latter time points. The day 1 fractions were prepared and analyzed immediately in triplicate after dissolution. The average concentration of the aged samples (triplicates) on each day was calculated and compared to the concentrations of day 1 samples. The percent change from the initial measurement was calculated for all aged samples and was found to be less than 5% after 14 days of storage in the above conditions.

3.2. Nicotine Release from on!® Pouches

Following method validation for oral nicotine pouches, we characterized the in vitro release profiles of nicotine from the 35 varieties of on!® nicotine pouch products. As an example of the release profiles measured, Figure 1 shows the cumulative release profiles (Figure 1A) and percent of total release (Figure 1B) of nicotine from on!® Mint pouches at five different nicotine levels (1.5, 2, 3.5, 4, and 8 mg per pouch). As expected, the cumulative nicotine released from the on!® pouches increases as the nicotine content of the product increases. The percent of total release profiles of nicotine from the on!® Mint pouches at various nicotine levels were equivalent (Figure 1B). More rapid nicotine dissolution was observed for all five products with a total percent release of ~80% in the profile region between zero and 20 min. The total percent of release for all products (>95%) was achieved within 40 min before the nicotine dissolution profiles reached a plateau. Despite differences in total nicotine content, on!® Mint pouches at various nicotine levels exhibit similar kinetic profiles. Similar observations were seen for the other flavor variants of on!® nicotine pouches including Citrus, Wintergreen, Coffee, Berry, Cinnamon, and Original across all five nicotine levels (Table 1).

To further confirm the above observations, we analyzed the nicotine release profiles by calculating the difference factor (f_1) and similarity factor (f_2) by adopting a methodology referenced in the Guidance for Industry from FDA's Center for Drug Evaluation and Research (CDER) [29,30]. Table 1 shows the f_1 and f_2 values obtained by using the 4 mg on!® nicotine pouches as the reference products for all flavor variants. In this study, we have chosen the on!® 4 mg as a comparator as it represents the mid-range nicotine concentration of all products. The f_1 and f_2 values for the 35 on!® nicotine pouches at different nicotine strengths and within each flavor variant demonstrate equivalency of the products with calculated f_1 lower than 15 and f_2 higher than 50. These data indicate that the total amount of nicotine content in on!® pouches does not affect the nicotine release profile.

To assess the influence of the flavor on the nicotine release rate, we evaluated the nicotine release profiles from all flavored on!® pouch products at each nicotine level. As an example, Figure 2 shows the cumulative release (Figure 2A) and percent of total release profiles (Figure 2B) of nicotine from the 3.5 mg on!® pouches with seven different flavor variants. The cumulative nicotine release profiles show that similar amounts of nicotine are released from the pouches (Figure 2B). The overlapping percent of total release profiles of nicotine indicate equivalency between the seven flavor variants of on!® pouch products at the 3.5 mg nicotine level. Moreover, the calculated f_1 and f_2 values demonstrated equivalency between these products (Table 2) using the on!® Mint nicotine pouches as a comparator. This shows that the flavor in the 3.5 mg on!® nicotine pouches do not influence the release profile of nicotine under our experimental conditions.

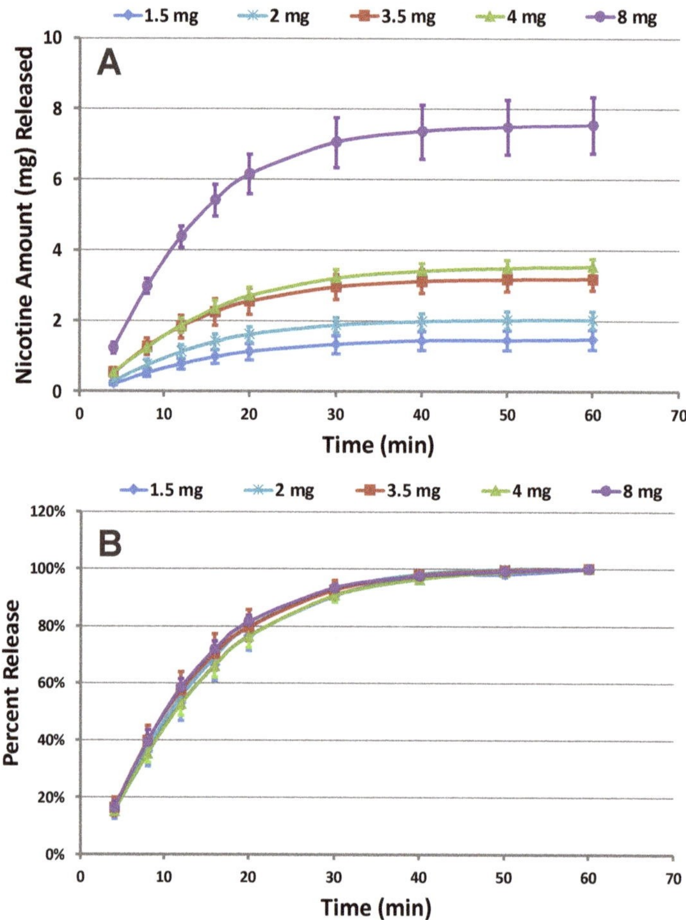

Figure 1. (**A**) Cumulative release and (**B**) percent of total dissolution release profiles of nicotine collected from Mint on!® pouches across all nicotine levels (n = 12) (Error Bars ± 1 SD).

Table 1. f_1 and f_2 values for on!® nicotine pouch comparisons across all nicotine levels for each flavor. The on!® 4 mg pouches for each flavor were used as the reference products for all comparisons.

	on!® Mint		
Compared Products	f_1	f_2	**Equivalency**
4 mg vs. 1.5 mg	0.8	97.4	Yes
4 mg vs. 2 mg	3.9	79.0	Yes
4 mg vs. 3.5 mg	5.9	71.5	Yes
4 mg vs. 8 mg	7.4	67.1	Yes

Table 1. *Cont.*

on!® Citrus			
Compared Products	**f_1**	**f_2**	**Equivalency**
4 mg vs. 1.5 mg	6.5	72.2	Yes
4 mg vs. 2 mg	5.7	71.3	Yes
4 mg vs. 3.5 mg	6.5	73.2	Yes
4 mg vs. 8 mg	6.3	72.8	Yes

on!® Wintergreen			
Compared Products	**f_1**	**f_2**	**Equivalency**
4 mg vs. 1.5 mg	1.1	94.3	Yes
4 mg vs. 2 mg	11.2	56.3	Yes
4 mg vs. 3.5 mg	4.8	72.8	Yes
4 mg vs. 8 mg	4.6	73.7	Yes

on!® Coffee			
Compared Products	**f_1**	**f_2**	**Equivalency**
4 mg vs. 1.5 mg	6.6	67.2	Yes
4 mg vs. 2 mg	4.5	74.8	Yes
4 mg vs. 3.5 mg	1.7	91.1	Yes
4 mg vs. 8 mg	2.6	86.1	Yes

on!® Berry			
Compared Products	**f_1**	**f_2**	**Equivalency**
4 mg vs. 1.5 mg	8.6	63.0	Yes
4 mg vs. 2 mg	4.1	76.4	Yes
4 mg vs. 3.5 mg	1.5	92.3	Yes
4 mg vs. 8 mg	6.4	67.6	Yes

on!® Cinnamon			
Compared Products	**f_1**	**f_2**	**Equivalency**
4 mg vs. 1.5 mg	8.1	63.8	Yes
4 mg vs. 2 mg	2.3	87.3	Yes
4 mg vs. 3.5 mg	2.3	86.9	Yes
4 mg vs. 8 mg	4.0	79.4	Yes

on!® Original			
Compared Products	**f_1**	**f_2**	**Equivalency**
4 mg vs. 1.5 mg	8.5	62.2	Yes
4 mg vs. 2 mg	4.0	77.3	Yes
4 mg vs. 3.5 mg	8.4	62.6	Yes
4 mg vs. 8 mg	3.0	82.8	Yes

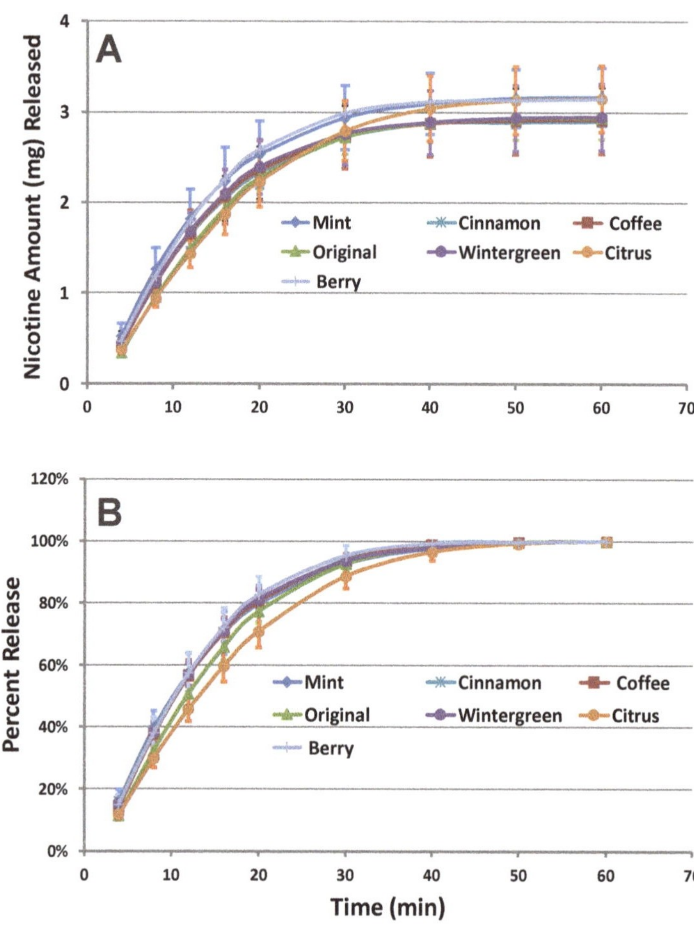

Figure 2. (**A**) Cumulative release and (**B**) percent of total dissolution release profiles of nicotine collected from all flavored on!® pouches at 3.5 mg nicotine level (n = 12) (Error Bars ± 1 SD).

Table 2. f_1 and f_2 values for on!® 3.5 mg nicotine pouch comparisons across all seven flavors. Mint on!® 3.5 mg nicotine pouches were used as a reference product for all comparisons.

Compared Flavors	f_1	f_2	Equivalency
Mint vs. Berry	3.0	82.3	Yes
Mint vs. Cinnamon	1.0	95.2	Yes
Mint vs. Coffee	1.9	88.9	Yes
Mint vs. Original	7.1	65.2	Yes
Mint vs. Wintergreen	1.7	91.5	Yes
Mint vs. Citrus	13.8	52.8	Yes

3.3. Comparison with Smokeless Tobacco and Other OTDN Pouch Products

To better understand the release rates of on!® nicotine pouches and how they compare to traditional smokeless tobacco products and other OTDN pouch products, we compared the nicotine release profiles of on!® nicotine pouches to Skoal® Bandits and Skoal®

pouches (commercially available traditional pouched smokeless tobacco products) and ZYN® nicotine pouches, another OTDN pouch products.

The on!® nicotine pouches are free of tobacco and do not have the same matrix content as the traditional pouched smokeless tobacco products. In addition, they are smaller in size and have a lower amount of nicotine per pouch compared to traditional pouched smokeless tobacco. Figure 3A shows the cumulative release profiles of nicotine from the Wintergreen flavored on!® 3.5 mg compared to Wintergreen flavored traditional smokeless tobacco pouch products, Skoal® Bandits and Skoal® pouches. The pouch weights for each product are 0.263 g, 0.72 g, and 1.55 g for on!®, Skoal® Bandits, and Skoal® pouches, respectively. The amount of total nicotine released (nicotine amount (mg) released) from Skoal® Bandits and Skoal® pouches as compared to the on!® product is attributed to the differences in nicotine concentration per pouch (Figure 3A). Despite the different pouch weight, nicotine concentration per pouch, and pouch composition, the percent nicotine released at each collection time point for the on!® and Skoal® Bandits pouches were found to be equivalent, as indicated by the overlapping release profiles (Figure 3B). However, the rate of nicotine release from the Skoal® pouches was found to be slower than on!® and Skoal® Bandits pouches (Figure 3B). In the profile region between zero and 20 min, a rapid dissolution was observed for on!® and Skoal® Bandits pouches, with a total percent release of 80% nicotine, whereas only 65% of the nicotine was released for Skoal® pouches. These observations were confirmed by calculating the f_1 and f_2 values. The calculated f_1 and f_2 values were 8.1 and 61.0 when comparing on!® to Skoal® Bandits, indicating equivalency between these products, and 21.1 and 46.0 when comparing on!® to Skoal® pouches, showing a difference in the nicotine release rates between these two products. These data illustrate that on!® nicotine pouches show similar or faster nicotine release profiles than the traditional pouched smokeless products tested here.

We also compared the performance of on!® to ZYN® pouches, a product marketed by Swedish Match North America. For this comparison, we selected the 3.5 mg and 8 mg on!® Mint, Wintergreen, and Coffee and 3 mg and 6 mg ZYN® Cool Mint, Wintergreen, and Coffee nicotine pouch products. As an example, Figure 4A shows the cumulative release profiles of nicotine from the 3.5 mg Mint on!® and 3 mg Cool Mint ZYN® pouches. Figure 4B shows the cumulative release profiles of nicotine from the 3.5 mg Wintergreen on!® and 3 mg Wintergreen ZYN® pouches. As expected, similar amounts of nicotine were released from both the on!® and the ZYN® products. The total release profiles of nicotine from the 3.5 mg on!® demonstrated a slightly slower release rate than the 3 mg ZYN® pouches (Figure 4C,D). However, when we calculate the f_1 and f_2 values, comparing on!® Mint to ZYN® Cool Mint, the nicotine release rate demonstrated equivalency, with f_1 and f_2 values of 9.2 and 60.1, respectively. In contrast, the on!® Wintergreen to ZYN® Wintergreen comparison resulted in a difference in the nicotine release rate with f_1 and f_2 values of 16.1 and 48.7, respectively. This observation was also seen when comparing other flavor variants and nicotine levels between on!® and ZYN® pouches (Table 3), indicating that the nicotine release profile and performance of these products could be similar but not necessarily statistically equivalent. Any measured and calculated differences between products within this OTDN category could be associated with inherent product characteristics (e.g., pouch paper and ingredients).

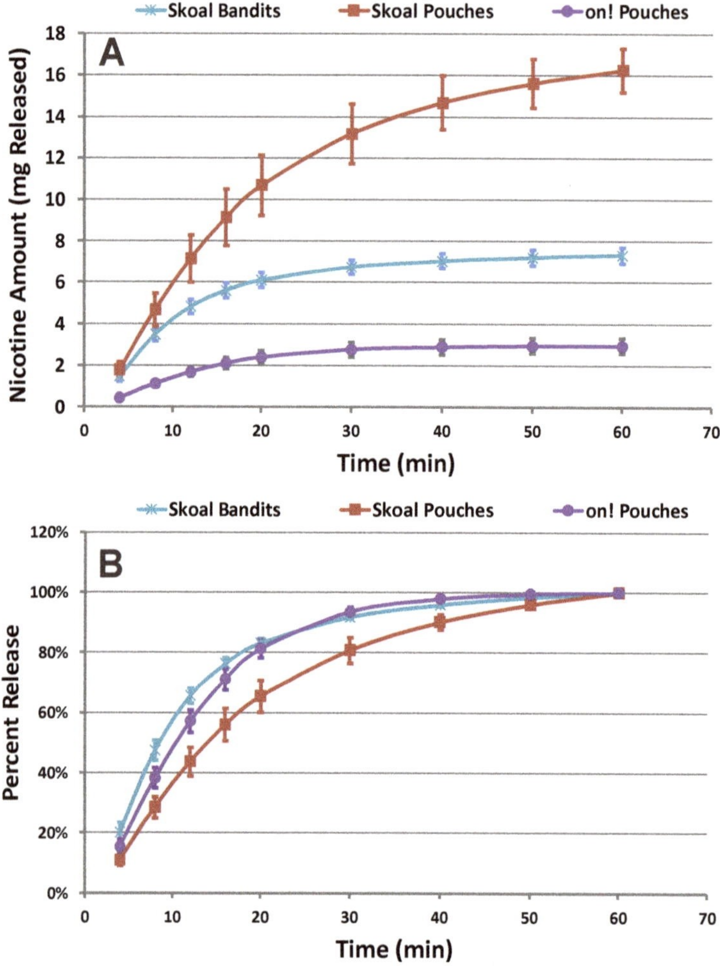

Figure 3. (**A**) Cumulative release and (**B**) percent of total dissolution release profiles of nicotine collected from Wintergreen flavored on!® 3.5 mg, Skoal® Bandits, and Skoal® pouches (n = 12) (Error Bars ± 1 SD).

Table 3. f_1 and f_2 values for on!® and ZYN® nicotine pouch comparisons.

Compared Flavors	f_1	f_2	Equivalency
on!® Mint 3.5 mg vs. ZYN® Cool Mint 3 mg	9.2	60.1	Yes
on!® Wintergreen 3.5 mg vs. ZYN® Wintergreen 3 mg	16.1	48.7	No
on!® Coffee 3.5 mg vs. ZYN® Coffee 3 mg	23.5	40.9	No
on!® Mint 8 mg vs. ZYN® Cool Mint 6 mg	12.6	52.8	Yes
on!® Wintergreen 8 mg vs. ZYN® Wintergreen 6 mg	18.9	44.9	No
on!®Coffee 8 mg vs. ZYN® Coffee 6 mg	9.1	61.1	Yes

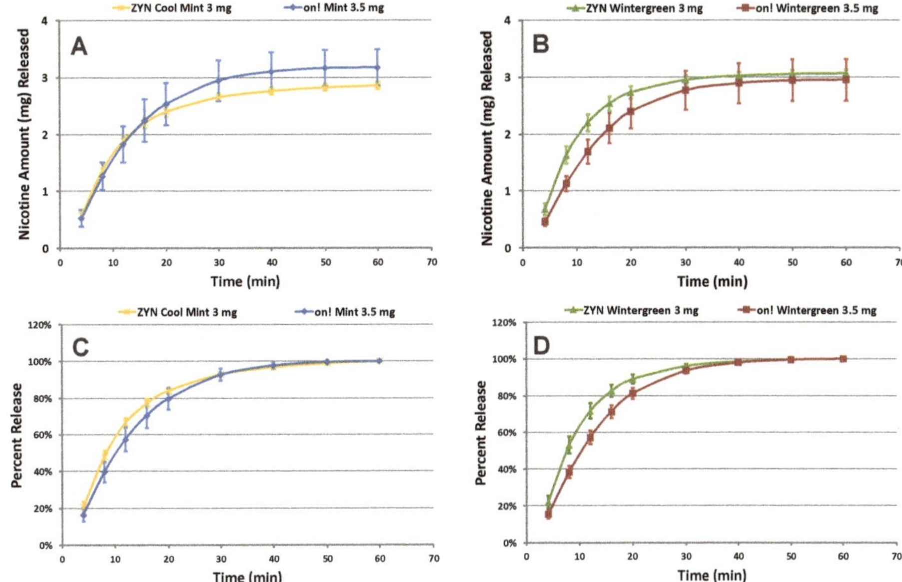

Figure 4. (**A,B**) Cumulative release and (**C,D**) percent of total dissolution release profiles of nicotine collected from Mint and Wintergreen on!® 3.5 mg and Cool Mint and Wintergreen ZYN® 3 mg pouches (n = 12) (Error Bars ± 1 SD).

4. Conclusions

In this report, we evaluated the release profile of nicotine from 35 on!® nicotine pouch products, which are currently marketed in seven flavor variants with five different nicotine levels. Our data show similar nicotine release profiles among the thirty-five (35) on!® products. Factor of difference (f_1) and factor of similarity (f_2) calculations confirmed similar product performance for all products. Nicotine release rate was not dependent on flavor and nicotine levels. Furthermore, we showed similarities and differences in the nicotine release rate from on!® nicotine pouches when compared to a few selected traditional pouched moist smokeless tobacco and non-traditional ZYN® pouch products. We believe that the data presented will provide useful information for product characterization and product-to-product comparisons. In addition, the dissolution data provided herein could be used to support clinical studies and establish future in vitro/in vivo (IVIV) correlations.

Author Contributions: Conceptualization, F.A.; validation, F.A., N.M. and C.S.; formal analysis, F.A.; investigation, F.A., N.M. and C.S.; resources, J.H.M.; data curation; writing—original draft preparation, F.A.; writing—review and editing, F.A., J.H.M. and T.L.D.; visualization, F.A.; supervision, F.A., J.H.M. and T.L.D.; project administration, F.A., T.L.D.; All authors have read and agreed to the published version of the manuscript.

Funding: This research received no external funding.

Informed Consent Statement: Informed consent was obtained from all subjects involved in the study.

Data Availability Statement: The data presented in this study are openly available in [repository name e.g., FigShare] at [doi], reference number [reference number].

Acknowledgments: The authors thank Mohamadi Sarkar, Jack Marshall, Christopher McFarlane, Karl Wagner, and James Skapars from Altria Client Services for the helpful discussions.

Conflicts of Interest: The authors declare no competing financial interest.

References

1. Mejia, A.B.; Ling, P.M.; Glantz, S.A. Quantifying the effects of promoting smokeless tobacco as a harm reduction strategy in the USA. *Tob. Control* **2010**, *19*, 297–305. [CrossRef] [PubMed]
2. Rodu, B.; Godshall, W.T. Tobacco harm reduction: An alternative cessation strategy for inveterate smokers. *Harm Reduct. J.* **2006**, *3*, 37. [CrossRef] [PubMed]
3. Hatsukami, D.K.; Lemmonds, C.; Tomar, S.L. Smokeless tobacco use: Harm reduction or induction approach? *Prev. Med.* **2004**, *38*, 309–317. [CrossRef] [PubMed]
4. Hatsukami, D.K.; Carroll, D.M. Tobacco harm reduction: Past history, current controversies and a proposed approach for the future. *Prev. Med.* **2020**, *140*, 106099. [CrossRef] [PubMed]
5. Kozlowski, L.T.; Sweanor, D.T. Young or adult users of multiple tobacco/nicotine products urgently need to be informed of meaningful differences in product risks. *Addict. Behav.* **2018**, *76*, 376–381. [CrossRef]
6. Centers for Disease Control and Preventions (CDC). Smokeless Tobacco: Products and Marketing. Available online: https://www.cdc.gov/tobacco/data_statistics/fact_sheets/smokeless/products_marketing/index.htm#not-burned (accessed on 15 October 2020).
7. Robichaud, M.O.; Seidenberg, A.B.; Byron, M.J. Tobacco companies introduce 'tobacco-free' nicotine pouches. *Tob Control* **2019**, *29*, e145–e146. [CrossRef]
8. Choi, J.H.; Dresler, C.M.; Norton, M.R.; Strahs, K.R. Pharmacokinetics of a nicotine polacrilex lozenge. *Nicotine Tob. Res.* **2003**, *5*, 635–644. [CrossRef]
9. West, R.; Shiffman, S. Effect of oral nicotine dosing forms on cigarette withdrawal symptoms and craving: A systematic review. *Psychopharmacology* **2001**, *155*, 115–122. [CrossRef]
10. O'Connor, R.J.; Norton, K.J.; Bansal-Travers, M.; Mahoney, M.C.; Cummings, K.M.; Borland, R. US smokers' reactions to a brief trial of oral nicotine products. *Harm Reduct. J.* **2011**, *8*, 1. [CrossRef]
11. Fisher, M.T.; Tan-Torres, S.M.; Gaworski, C.L.; Black, R.A.; Sarkar, M.A. Smokeless tobacco mortality risks: An analysis of two contemporary nationally representative longitudinal mortality studies. *Harm Reduct. J.* **2019**, *16*, 27. [CrossRef]
12. Gottlieb, S.; Zeller, M. A Nicotine-Focused Framework for Public Health. *N. Engl. J. Med.* **2017**, *377*, 1111–1114. [CrossRef] [PubMed]
13. National Institute for Healthcare and Excellence (NICE). Smoking: Harm Reduction. 2013. Available online: https://www.nice.org.uk/guidance/ph45/resources/smoking-harm-reduction-pdf-1996359619525 (accessed on 15 October 2020).
14. Hukkanen, J.; Jacob, P.; Benowitz, N.L. Metabolism and Disposition Kinetics of Nicotine. *Pharmacol. Rev.* **2005**, *57*, 79. [CrossRef]
15. Qureshi, S.A. In vitro-in vivo correlation (ivivc) and determining drug concentrations in blood from dissolution testing–a simple and practical approach. *Open Drug Deliv. J.* **2010**, *4*, 38–47. [CrossRef]
16. Williams, R.L.; Foster, T.S. Dissolution; a continuing perspective. *Dissolution Technol. Augysr.* **2004**, *6*, 14. [CrossRef]
17. Wang, Q.; Fotaki, N.; Mao, Y. Biorelevant dissolution: Methodology and application in drug development. *Dissolution Technol.* **2009**, *16*, 6–12. [CrossRef]
18. USP. 711 Dissolution USP. Available online: https://www.usp.org/sites/default/files/usp/document/harmonization/gen-method/q01_pf_ira_33_4_2007.pdf (accessed on 15 October 2020).
19. Zieschang, L.; Klein, M.; Krämer, J.; Windbergs, M. In Vitro Performance Testing of Medicated Chewing Gums. *Dissolution Technol.* **2018**, *25*, 64–69. [CrossRef]
20. Dressman, J.B.; Reppas, C. In vitro–in vivo correlations for lipophilic, poorly water-soluble drugs. *Eur. J. Pharm. Sci.* **2000**, *11*, S73–S80. [CrossRef]
21. Amidon, G.L.; Lennernäs, H.; Shah, V.P.; Crison, J.R. A theoretical basis for a biopharmaceutic drug classification: The correlation of in vitro drug product dissolution and in vivo bioavailability. *Pharm. Res.* **1995**, *12*, 413–420. [CrossRef]
22. Morjaria, Y.; Irwin, W.J.; Barnett, P.X.; Chan, R.S.; Conway, B.R. In Vitro Release of Nicotine From Chewing Gum Formulations. *Dissolution Technol.* **2004**, *11*, 12–15. [CrossRef]
23. Delvadia, P.R.; Barr, W.H.; Karnes, H.T. A biorelevant in vitro release/permeation system for oral transmucosal dosage forms. *Int. J. Pharm.* **2012**, *430*, 104–113. [CrossRef]
24. Nasr, M.M.; Reepmeyer, J.C.; Tang, Y. In vitro study of nicotine release from smokeless tobacco. *J. AOAC Int.* **1998**, *81*, 540–543. [CrossRef] [PubMed]
25. Li, P.; Zhang, J.; Sun, S.-H.; Xie, J.-P.; Zong, Y.-L. A novel model mouth system for evaluation of In Vitro release of nicotine from moist snuff. *Chem. Cent. J.* **2013**, *7*, 176. [CrossRef] [PubMed]
26. Cecil, T.L.; Brewer, T.M.; Holman, M.; Ashley, D.L. Food and Drug Administration. Dissolution as a Critical Comparison of Smokeless Product Performance: SE Requirements and Recommendations for the Review of Dissolution Studies. Memorandum from Cecil. 2016. Available online: https://www.fda.gov/media/124673/download (accessed on 15 October 2020).
27. Kvist, C.; Andersson, S.B.; Fors, S.; Wennergren, B.; Berglund, J. Apparatus for studying in vitro drug release from medicated chewing gums. *Int. J. Pharm.* **1999**, *189*, 57–65. [CrossRef]
28. Miller, J.H.; Danielson, T.; Pithawalla, Y.B.; Brown, A.P.; Wilkinson, C.; Wagner, K.; Aldeek, F. Method development and validation of dissolution testing for nicotine release from smokeless tobacco products using flow-through cell apparatus and UPLC-PDA. *J. Chromatogr. B Anal. Technol. Biomed. Life Sci.* **2020**, *1141*, 122012. [CrossRef]

29. Food and Drug Administration. SUPAC-IR: Immediate-Release Solid Oral Dosage Forms: Scale-Up and Post-Approval Changes: Chemistry, Manufacturing and Controls, In Vitro Dissolution Testing, and In Vivo Bioequivalence Documentation. 1995. Available online: https://www.fda.gov/regulatory-information/search-fda-guidance-documents/supac-ir-immediate-release-solid-oral-dosage-forms-scale-and-post-approval-changes-chemistry (accessed on 15 October 2020).
30. Food and Drug Administration. Guidance for Industry: Dissolution Testing of Immediate Release Solid Oral Dosage Forms. 1997. Available online: https://www.fda.gov/regulatory-information/search-fda-guidance-documents/dissolution-testing-immediate-release-solid-oral-dosage-forms (accessed on 15 October 2020).
31. DIN V Test Method 53160-1. Colorfastness to Saliva; Determination of the Colorfastness of Articles in Common Use Part 1: Resistance to Artificial Saliva. Available online: http://www.manufacturingsolutionscenter.org/colorfastness-to-saliva-testing.html (accessed on 15 October 2020).
32. Shah, V.P.; Tsong, Y.; Sathe, P.; Liu, J.-P. In Vitro Dissolution Profile Comparison—Statistics and Analysis of the Similarity Factor, f2. *Pharm. Res.* **1998**, *15*, 889–896. [CrossRef]
33. Chow, S.C.; Ki, F.Y. Statistical comparison between dissolution profiles of drug products. *J. Biopharm. Stat.* **1997**, *7*, 241–258. [CrossRef] [PubMed]
34. Kapoor, D.; Maheshwari, R.; Verma, K.; Sharma, S.; Pethe, A.; Tekade, R.K. Chapter 1—Fundamentals of diffusion and dissolution: Dissolution testing of pharmaceuticals. In *Drug Delivery Systems*; Tekade, R.K., Ed.; Academic Press: Cambridge, MA, USA, 2020; pp. 1–45.
35. Long, M.; Chen, Y. Chapter 14—Dissolution Testing of Solid Products. In *Developing Solid Oral Dosage Forms*; Qiu, Y., Chen, Y., Zhang, G.G.Z., Liu, L., Porter, W.R., Eds.; Academic Press: San Diego, CA, USA, 2009; pp. 319–340.

 *separations*

Article

Introducing a Novel Biorelevant In Vitro Dissolution Method for the Assessment of Nicotine Release from Oral Tobacco-Derived Nicotine (OTDN) and Snus Products

Matthias M. Knopp [1,*], Nikolai K. Kiil-Nielsen [1], Anna E. Masser [2] and Mikael Staaf [2]

[1] Bioneer:FARMA, Department of Pharmacy, Universitetsparken 2, DK-2100 Copenhagen, Denmark; nkn@bioneer.dk
[2] Swedish Match North Europe AB, Regulatory & Scientific Affairs, P.O. Box 17037, S-104 62 Stockholm, Sweden; anna.masser@swedishmatch.com (A.E.M.); mikael.staaf@swedishmatch.com (M.S.)
* Correspondence: mmk@bioneer.dk

Abstract: The rate at which oral tobacco-derived nicotine (OTDN) and snus pouches release nicotine into saliva is crucial to determine product performance. As no standardized method is available for this purpose, this study sought to develop a biorelevant dissolution method that could both discriminate between different products and predict in vivo behavior. Using a μDISS Profiler™ as a surrogate for the US Pharmacopoeia standard apparatuses and a custom-made sinker, nicotine release from an OTDN pouch product (ZYN® Dry Smooth) and a snus product (General® Pouched Snus White Portion Large) was determined in biorelevant volumes (10 mL) of artificial saliva. In addition, nicotine extraction in vivo was measured for both products. Strikingly, the method showed distinct dissolution curves for OTDN and snus pouches, and the nicotine release observed in vitro did not significantly differ from the nicotine extracted in vivo. The custom-made sinker was designed to accommodate both loose and pouched oral tobacco/nicotine products, and thus the proposed in vitro dissolution method is suitable to assess nicotine release from OTDN and snus pouches. Apart from providing individual dissolution curves, the method was also able to predict in vivo nicotine extraction. Thus, this method could serve as a (biorelevant) monograph for product equivalence studies.

Keywords: oral tobacco derived nicotine (OTDN) pouches; snus; nicotine release; nicotine dissolution; nicotine extraction; equivalence

Citation: Knopp, M.M.; Kiil-Nielsen, N.K.; Masser, A.E.; Staaf, M. Introducing a Novel Biorelevant In Vitro Dissolution Method for the Assessment of Nicotine Release from Oral Tobacco-Derived Nicotine (OTDN) and Snus Products. *Separations* **2022**, *9*, 52. https://doi.org/10.3390/separations9020052

Academic Editor: Alena Kubatova

Received: 10 January 2022
Accepted: 13 February 2022
Published: 15 February 2022

Publisher's Note: MDPI stays neutral with regard to jurisdictional claims in published maps and institutional affiliations.

1. Introduction

Oral tobacco-derived nicotine (OTDN) pouch products are growing in popularity, but cigarettes remain the most common tobacco product worldwide [1]. Cigarettes are a huge health burden in terms of tobacco-related morbidity and mortality, and smoking tobacco is the factor contributing to most preventable deaths worldwide [2]. The hazards of cigarettes are a result of the inhaled combustion products formed during smoking and have less to do with their tobacco and nicotine content [3]. Still, nicotine is the addictive substance sustaining cigarette dependence. Smokeless tobacco products (STPs) expose the user to no combustion products and epidemiological data on the STP Swedish snus have shown the use of Swedish snus to be significantly less harmful, in terms of morbidity and mortality, compared to cigarettes [4–6]. Little data are available on OTDN pouches, but current literature indicate that they contain less (potentially) harmful constituents and are less toxic in vitro, compared to cigarettes [7,8]. OTDN pouches are therefore an enticing alternative to traditional tobacco-based products, in terms of harm reduction.

OTDN pouches come in small, white sachets that are intended to be placed between the gum and upper lip where nicotine is released into the surrounding saliva from which it permeates the buccal mucosa, and subsequently enters systemic circulation. Thus, the performance of these products depends on the rate at which nicotine is released. However, despite growing attention among regulatory agencies and tobacco researchers, a standardized method to evaluate the nicotine release from OTDN pouches is yet to be established.

In vitro, nicotine release can be measured by dissolution testing, a method commonly used for pharmaceuticals. Therefore, it would be logical to glance at pharmaceutical guidance's and monographs when developing a novel dissolution method. In the United States (US), STPs and OTDN pouches are regulated by the Food and Drug Administration (FDA) Center for Tobacco Products, which requires that dissolution testing is carried out on novel nicotine products as well as to demonstrate product equivalence [9].

For STPs, such as snus and moist snuff, only limited literature on development of dissolution methods that are designed to discriminate between OTDN products or simulate in vivo nicotine release is available [10–15]. One of these studies utilizing the US Pharmacopeia type 4 apparatus (USP-4) (flow-through cell) method was able to discriminate between moist snuff and OTDN pouches [15]. However, this method saw a 77% nicotine release from a Swedish-style snus pouch after 30 min, which greatly differs from the in vivo situation where only 31–46% nicotine extraction from the same product is reported after 1 h [13,16–18]. As the amount of agitation on the products in the USP-4 is minimal, this inconsistency could be due to the amount of flow/volume of artificial saliva used in this method (4 mL/min) which is almost 10-fold higher than the unstimulated saliva flow rate (0.5 mL/min) [19].

Other methods utilizing the USP-1 and USP-2 (basket and paddle, respectively) were able to discriminate between the dissolution curves for moist snuff and Swedish snus. The authors pointed out the benefits of using USP-1 (and USP-2) being that they are the most used apparatuses, come at a lower cost than USP-4 and their ease of use [20]. Slower rates of nicotine release were shown, although still significantly faster than in vivo, probably also due to the large amount (500 mL) of artificial saliva used. Moreover, it is unclear if these methods could also discriminate between moist snuff/snus and OTDN pouches.

In this study, the in vitro release of nicotine from the OTDN pouch product ZYN® Dry Smooth and General® Pouched Snus White Portion Large (PSWL) was investigated in biorelevant volumes of artificial saliva using a µDISS Profiler™ dissolution method. The in vitro release data (i.e., biorelevance of the proposed dissolution method) was verified through in vivo nicotine extraction studies on the same products.

2. Materials and Methods

2.1. Investigational Products, Standards, and Reagents

General PSWL is a Swedish snus product containing 8 mg of nicotine. It comes in a rectangular pouch measuring 18 × 33 mm that weighs 1.0 g with a moisture content of 53.5% and a pH of 8.7. The pouch contains ground, air-cured tobacco, water, sodium chloride, sodium carbonate, humidifying agents, and food-grade flavorings.

ZYN Dry Smooth is an OTDN pouch product containing 6 mg of nicotine. The pouch measures 14 × 28 mm, weighs 0.4 g, contains 3% moisture, and has a pH of 8.3. The pouch contains fillers (maltitol and microcrystalline cellulose), a stabilizer (hydroxypropyl cellulose), pH adjusters (sodium carbonate and sodium bicarbonate), nicotine salt, food-grade flavorings, and a sweetener (acesulfame K).

Both products were provided by the manufacturer Swedish Match North Europe AB.

2.2. Standards and Reagents

A nicotine reference standard (>99.9%) was sourced from Łukasiewicz IPO (Warsaw, Poland). Saliva Orthana® (artificial saliva) containing (per 100 mL aqueous solution): porcine gastric mucin 3500 mg, methyl-4-hydroxybenzoate 100 mg, benzalkonium chloride 2 mg, EDTA disodium salt. H_2O (E386) 50 mg, H_2O_2 250 ppm, xylitol 2000 mg, peppermint oil 5 mg, spearmint oil 5 mg, NaCl 45 mg, KCl 63 mg, $CaCl_2$ 30 mg, K_2HPO_4 10 mg, KOH 76 mg with a neutral pH, was purchased from Biofac A/S (Kastrup, Denmark).

2.3. Sinker Preparation

As there is no suitable sinker commercially available, a custom-made sinker was prepared by 3D printing a 12 mm tall, hollow tube with an outside diameter of 21 mm and a wall thickness of 2 mm using polylactic acid (PLA). The structure was designed by computer aided design (CAD) using the online platform Tinkercad from Autodesk (San Rafael, CA, USA), exported as .stl files and converted to a readable file for the printer using the Cura software (version 3.6.0) from Ultimaker (Geldermase, The Netherlands). An Ultimaker 3 extended from Ultimaker (Geldermalsen, The Netherlands) was used to print the tube structures from a 2.85 mm PLA 3D printer filament (Innofil3D BV, Emmen, Netherlands) using a printing temperature of 200 °C and a layer height of 50 µm. After printing, the tube was fitted with a 20-mesh stainless steel sieve (0.84 mm sieve opening) by molding it into the PLA tube at 200 °C using a hot-plate. A picture of the final product loaded with an OTDN pouch can be seen in Figure 1A.

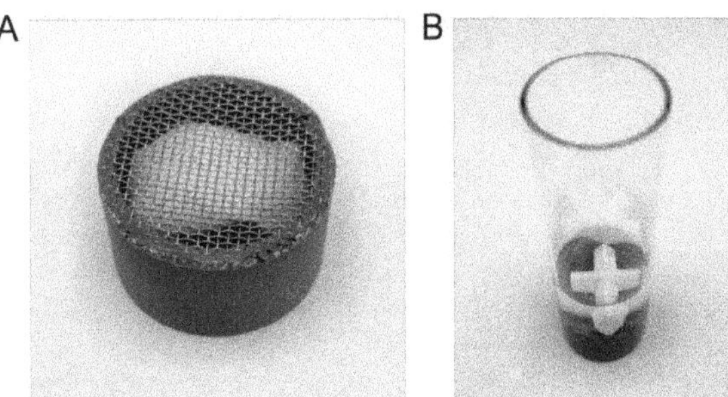

Figure 1. Details of the µDISS Profiler™ experimental setup. (**A**) An OTDN pouch loaded in the custom-made sinker, 3D printed using polylactic acid (PLA) and fitted with a 20-mesh stainless steel sieve. The sinker measures 12 mm in height and 21 mm in width. (**B**) A snus pouch loaded in the custom-made sinker and placed on the bottom of the standard 20 mL µDISS Profiler™ dissolution vessel containing 10 mL of artificial saliva with a 20 mm cross-shaped magnetic stirrer on top.

2.4. In Vitro Nicotine Release

The nicotine release experiments were carried out using a µDISS Profiler™ (Pion Inc., Billerica, MA, USA) with six channels without the in-line UV probes connected. The snus/OTDN pouches were weighed individually, loaded in the sinkers, and placed on the bottom (mesh up) of the standard 20 mL dissolution vessels with 20 mm cross-shaped magnetic stirrers on top. A picture of the experimental setup is shown in Figure 1B. The magnetic stirrers were set to operate at 100 rpm and the minibath temperature was set to 37 °C. The experiment was initiated by addition of 10 mL of artificial saliva (preheated to 37 °C) to each of the six dissolution vessels. Samples of 250 µL were taken at t = 5, 10, 15, 30, 45, 60, 90, and 120 min and replaced with 250 µL of preheated artificial saliva. The samples were diluted immediately with 375 µL acetonitrile and 375 µL ethanol in a 1.5 mL Eppendorf centrifuge tube to precipitate proteins from the saliva and avoid potential precipitation of the nicotine upon cooling. Diluted samples were centrifuged at 10,000 rpm for 10 min at room temperature and the resulting supernatant was analyzed for nicotine content using high-performance liquid chromatography (HPLC). The dissolution testing was performed on 12 dosage units of each formulation in accordance with regulatory guidelines [21].

2.5. Quantitative Analysis

Nicotine released in vitro was quantified by HPLC-UV using an Ultimate 3000 HPLC system (Dionex, Sunnyvale, CA, USA). A reverse phase Kinetex Evo C18 100A column (4.6 × 100 mm, 2.6 µm) (Phenomenex, Værløse, Denmark) was used for the separation and the mobile phases consisted of (A) 15 mM ammonium formate adjusted to pH 10.5 using triethylamine and (B) acetonitrile, which were pumped isocratically at 75% A and 25% B. A volume of 10 µL was injected and eluted at a flow rate of 0.5 mL/min, and the effluent was detected at a wavelength of 260 nm with a retention time of approximately 4.3 min. The concentration of nicotine in the samples was calculated using the mean value of the peak areas obtained from a calibration standard curve prepared in triplicate. Representative chromatograms are shown in Supplemental Figure S1. The method was validated for linearity, accuracy (recovery), range, precision (repeatability), limit of detection (LOD) and limit of quantification (LOQ) prior to use. The triplicate standard curve for nicotine was linear with an $r^2 = 0.9998$ over the range 1.95–500 µg/mL and a y-intercept at 0.86% of the target concentration response (200 µg/mL). At 50%, 100%, and 150% of the target concentration response, the recovery of nicotine was 98.95–99.62%. The precision of the retention time, peak area and peak height for nicotine was 0.12–0.50%. The LOD and LOQ for nicotine was 0.23 µg/mL and 0.75 µg/mL, respectively.

2.6. In Vivo Nicotine Extraction

A non-blinded, crossover, single-dose administration study was conducted to obtain the in vivo nicotine extraction data [22]. The study enrolled healthy male and female snus users aged ≥19 years, willing and able to give written informed consent. The study was carried out in accordance with ethical principles that have their origin in the Declaration of Helsinki and are consistent with International Council for Harmonization (ICH)/Good Clinical Practice (GCP), European Union Clinical Trials Directive, and applicable local regulatory requirements. The study was approved by the Swedish Ethical Review Authority and registered on the ISRCTN registry (ISRCTN44913332). The 18 subjects kept the pouch still between the upper lip and gum. Each used pouch, 18 replicates per product and time point (15 and 60 min), was collected and frozen ($-20\,°C$) pending nicotine analysis. Unused pouches, 10 replicates per product, were collected and frozen ($-20\,°C$) for analysis as references in the calculations of extracted doses. The concentration of nicotine in pouches was determined using a Gas Chromatography/Mass Spectrometry (GC/MS) system (Agilent 7890A GC, 7693A autosampler and 5975C MS) using an Agilent Innowax, 60 m × 0.25 mm ID column with a 0.25 µm film.

2.7. Data Analysis

Results from the in vitro and in vivo studies are expressed as mean ± standard deviation (SD). Statistical analysis was performed in SigmaPlot 14.0 from Systat Software Inc. (Chicago, IL, USA). A Student's t-test was performed on untransformed data to identify significant differences between in vitro and in vivo nicotine release after 15 min and 60 min.

A mathematical approach recognized by the US FDA was used to compare the similarities and differences in dissolution profiles [21,23]. The difference factor (f_1) was used to calculate the percent difference between two curves at each time point, which measures the absolute relative error between the two points. The similarity factor (f_2) measures the similarity in the percent dissolution between two curves. The two factors were calculated using the following equations:

$$f_1 = \left(\frac{\sum_{t=1}^{n} |R_t - T_t|}{\sum_{t=1}^{n} R_t} \right) * 100$$

$$f_2 = 50 * log \left(100 * \left(1 + \frac{\sum_{t=1}^{n} (R_t - T_t)^2}{n} \right)^{-\frac{1}{2}} \right)$$

R_t and T_t are the cumulative percentage dissolved of reference product and test product at time t, respectively, and n is the number of timepoints. Curves are considered similar for f_1 values close to 0, and f_2 values close to 100. Generally, dissolution profiles are judged to be equivalent if f_1 values are below 15 and if f_2 values are greater than 50.

3. Results and Discussion

Several different in vitro nicotine release methods have been proposed that are able to discriminate between different STPs, but common for these methods is that they all seem to substantially overpredict the nicotine release in vivo [13,20]. Therefore, this work sought to develop a discriminative and biorelevant in vitro method for nicotine release/dissolution from OTDN pouches and snus products and validate this with in vivo nicotine extraction data. To determine what volumes are biorelevant a reasoning made by the FDA in their memorandum regarding dissolution testing were followed [9]. There, usage time and salivary flow at the site where the pouch is placed were taken into consideration. No data on average usage time for ZYN Dry are available, but for pouched snus 65 min has been reported [24]. Both ZYN Dry and snus are placed under the upper lip where the parotid glands secrete saliva into the mouth. The resting and stimulated flow rate of saliva from the parotid glands are 0.1 mL/min and 1.05 mL/min, respectively [25,26]. Assuming that the pouch is kept for 60 min, the average flow of saliva from both parotid glands would vary between 6–63 mL, depending on degree of stimuli. As the pouch is kept on one side of the mouth, and therefore is mainly in contact with saliva from one of the glands, 3–33 mL can be considered biorelevant. Here, 10 mL of test medium was chosen, and the µDISS Profiler™ was used as a surrogate for the USP standard dissolution apparatuses, to allow for the low volume. As there is no standard simulated saliva fluid recipe described in the US Pharmacopeia, we chose a commercially available artificial saliva as dissolution medium as opposed to less viscous buffer systems. Finally, a custom-made sinker was designed to prevent floating and pouch/material discharge during the experiments, and to accommodate both loose and pouched products (Figure 1A).

Using the novel dissolution method, nicotine release profiles for ZYN Dry Smooth and General PSWL were obtained (Figure 2). The average nicotine release profiles are plotted as percentage of dose, to account for differences in nicotine dose due to pouch filling/weight variance. After 15 min and 60 min, the nicotine release from General PSWL was $9.2 \pm 4.7\%$ and $29.9 \pm 11.2\%$, respectively. For ZYN Dry Smooth, the nicotine release after 15 min and 60 min, was $15.3 \pm 7.2\%$ and $50.1 \pm 14.5\%$, respectively.

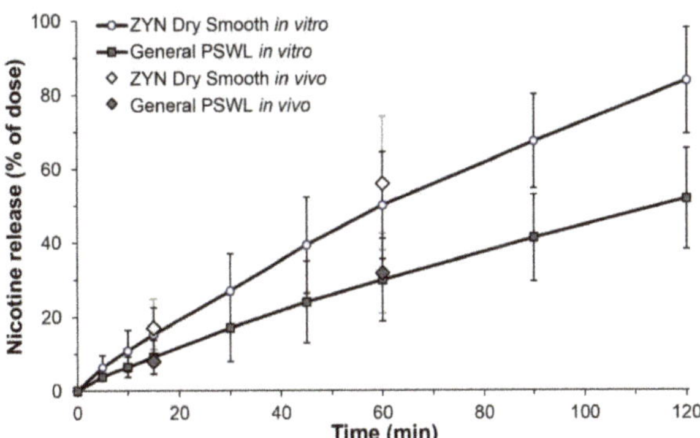

Figure 2. In vitro and in vivo nicotine release from ZYN Dry Smooth and General Pouched Snus White Portion Large (PSWL) as a function of time. Nicotine in vitro release profiles from General PSWL (dark gray squares, black error bars) and ZYN Dry Smooth (white circles, black error bars) in artificial saliva as % of dose ± standard deviation (SD) ($n = 12$). Based on calculations of the difference (f_1) and similarity factor (f_2) the curves are distinct ($f_1 = 63.6$, $f_2 = 38.8$). In vivo nicotine extraction after 15 min and 60 min for ZYN Dry Smooth (white diamonds, grey error bars) and General PSWL (dark grey diamonds, grey error bars) is added for comparison as % of dose ± SD ($n = 18$). No significant differences were seen at 15 and 60 min between in vitro and in vivo conditions for both products, respectively.

To investigate if the in vitro dissolution method reflected on actual in vivo conditions the results were compared to a previously conducted clinical study. The study enrolled 18 daily snus users aged $\geq$19 years which kept one pouch at a time still between the upper lip and gum. The in vivo extracted fraction of nicotine from General PSWL, after 15 and 60 min, was 8.0 $\pm$ 3.3% and 31.8 $\pm$ 10.8%, respectively. For ZYN Dry Smooth, the in vivo extracted fraction of nicotine, after 15 and 60 min, was 17.1 $\pm$ 7.8% and 56.0 $\pm$ 18.1%, respectively (Figure 2). General PSWL contains 33% more nicotine per pouch than ZYN Dry Smooth does. However, the absolute nicotine release was 32% higher in ZYN Dry Smooth because of a higher extracted fraction of nicotine. The results are in line with previously published data on Swedish snus and ZYN Dry Smooth [16–18,27]. The large difference in extracted fractions between both products are likely an effect of their diverse characteristics, in terms of nicotine source (ground tobacco leaves vs. nicotine salt), moisture, pH, and pouch geometry.

A critical feature of any dissolution method is that it should be able to distinguish between different products, here an OTDN and a snus pouch product. To compare the similarities and differences in dissolution profiles a mathematical approach recognized by the US FDA were used [21,23]. Based on this method, curves are considered similar for f_1 values close to 0, and f_2 values close to 100. In contrast, dissolution profiles are judged to be distinct if f_1 values are above 15 and if f_2 values are smaller than 50. We obtained f_1 and f_2 values of 63.6 and 38.8, respectively, showing that the curves for ZYN Dry Smooth and General PSWL are distinct.

A second feature of dissolution testing is that it can be used to predict the in vivo behavior of a product. A Student's t-test was used to verify similarities and differences between in vitro and in vivo conditions. Strikingly, no significant differences ($p > 0.05$) were seen for both products, at the two time points tested. As earlier studies have used much higher media volumes [13,20], it seems that the volume of saliva is of great importance when comparing nicotine release from OTDN and snus products.

In summary, this indicates that not only is the proposed in vitro nicotine release method able to discriminate between products from two different product categories, but it is also predictive of in vivo nicotine release, at least for the products tested. Thus, this method could serve as a predictive tool for product development and/or a monograph for oral tobacco/nicotine product equivalence studies.

4. Conclusions

In this work, a novel dissolution method was developed and the nicotine extraction from an OTDN pouch product (ZYN Dry Smooth) and a snus product (General PSWL) was determined. Calculations of the difference and similarity factor showed distinct nicotine-release curves for the two different products, verifying that the method can discriminate between different product categories. To investigate if the in vitro method could predict in vivo behavior, in vivo nicotine extraction was measured for both products and both time points. No significant differences could be seen within products when comparing in vitro and in vivo data after 15 min and 60 min.

Consequently, this method is to the best of our knowledge the first method developed that is both sensitive enough to discriminate between a product containing purified nicotine (ZYN Dry Smooth) and a product containing tobacco (General PSWL), as well as to be able to predict in vivo behavior.

Finally, the custom-made sinker was designed to accommodate both loose and pouched snus/snuff material. Thus, the proposed in vitro dissolution method could potentially be applied to assess the nicotine release from other oral nicotine/tobacco products e.g., moist snuff, dry snuff, and dissolvables.

Supplementary Materials: The following supporting information can be downloaded at: https://www.mdpi.com/article/10.3390/separations9020052/s1.

Author Contributions: Conceptualization, M.M.K. and M.S.; methodology, M.M.K.; validation, A.E.M., N.K.K.-N., M.M.K. and M.S.; formal analysis, M.M.K. and N.K.K.-N.; writing—original draft preparation, A.E.M. and M.M.K.; writing—review and editing, A.E.M., M.M.K., M.S. and N.K.K.-N.; visualization, M.M.K. All authors have read and agreed to the published version of the manuscript.

Funding: This research received no external funding.

Institutional Review Board Statement: The study was conducted in accordance with the Declaration of Helsinki, and approved by the Swedish Ethical Review Authority (protocol code Dnr 2017/318 approved 13 September 2017).

Informed Consent Statement: Informed consent was obtained from all subjects involved in the study.

Data Availability Statement: The data presented in this study are available on request from the corresponding author. The data are not publicly available due to company policies.

Acknowledgments: The authors would like to thank R. Pendrill for valuable help with f_1 and f_2 calculations.

Conflicts of Interest: A.E.M. and M.S. are employees of Swedish Match North Europe AB who markets the ZYN and general products described in the article. M.M.K. has received funding from Swedish Match North Europe AB.

References

1. Delnevo, C.D.; Hrywna, M.; Miller Lo, E.J.; Wackowski, O.A. Examining market trends in smokeless tobacco sales in the United States: 2011–2019. *Nicotine Tob. Res.* **2021**, *23*, 1420–1424. [CrossRef]
2. The World Health Organization. WHO Global Report: Mortality Attributable to Tobacco. Available online: http://apps.who.int/iris/bitstream/handle/10665/44815/9789241564434_eng.pdf (accessed on 2 June 2021).
3. Institute of Medicine. *Clearing the Smoke: Assessing the Science Base for Tobacco Harm Reduction*; The National Academies Press: Washington, DC, USA, 2001. [CrossRef]
4. Ramström, L.; Wikmans, T. Mortality attributable to tobacco among men in Sweden and other European countries: An analysis of data in a WHO report. *Tob. Induc. Dis.* **2014**, *12*, 14. [CrossRef]
5. Foulds, J.; Ramstrom, L.; Burke, M.; Fagerström, K. Effect of smokeless tobacco (snus) on smoking and public health in Sweden. *Tob. Control* **2003**, *12*, 349–359. [CrossRef]
6. Lee, P.N. Epidemiological evidence relating snus to health—An updated review based on recent publications. *Harm Reduct. J.* **2013**, *10*, 36. [CrossRef]
7. Azzopardi, D.; Liu, C.; Murphy, J. Chemical characterization of tobacco-free "modern" oral nicotine pouches and their position on the toxicant and risk continuums. *Drug Chem. Toxicol.* **2021**, 1–9. [CrossRef]
8. Bishop, E.; East, N.; Bozhilova, S.; Santopietro, S.; Smart, D.; Taylor, M.; Meredith, S.; Baxter, A.; Breheny, D.; Thorne, D.; et al. An approach for the extract generation and toxicological assessment of tobacco-free 'modern' oral nicotine pouches. *Food Chem. Toxicol.* **2020**, *145*, 111713. [CrossRef]
9. Cecil, T.L.; Brewer, T.M.; Holman, M.; Ashley, D.L. Dissolution as a Critical Comparison of Smokeless Product Performance: SE Requirements and Recommendations for the Review of Dissolution Studies. Memorandum from Cecil. Available online: https://www.fda.gov/media/124673/download (accessed on 10 September 2021).
10. Nasr, M.M.; Reepmeyer, J.C.; Tang, Y. In vitro study of nicotine release from smokeless tobacco. *J. AOAC Int.* **1998**, *81*, 540–543. [CrossRef]
11. Delvadia, P.R.; Barr, W.H.; Karnes, H.T. A biorelevant in vitro release/permeation system for oral transmucosal dosage forms. *Int. J. Pharm.* **2012**, *430*, 104–113. [CrossRef]
12. Li, P.; Zhang, J.; Sun, S.H.; Xie, J.P.; Zong, Y.L. A novel model mouth system for evaluation of In Vitro release of nicotine from moist snuff. *Chem. Cent. J.* **2013**, *7*, 176. [CrossRef] [PubMed]
13. Miller, J.H.; Danielson, T.; Pithawalla, Y.B.; Brown, A.P.; Wilkinson, C.; Wagner, K.; Aldeek, F. Method development and validation of dissolution testing for nicotine release from smokeless tobacco products using flow-through cell apparatus and UPLC-PDA. *J. Chromatogr. B Anal. Technol. Biomed. Life Sci.* **2020**, *1141*, 122012. [CrossRef]
14. Rahman, Z.; Mohamed, E.M.; Dharani, S.; Khuroo, T.; Young, M.; Feng, C.; Cecil, T.; Khan, M.A. Development and Validation of a Discriminatory Dissolution Method for Portioned Moist Snuff and Snus. *J. Pharm. Sci.* **2021**. [CrossRef] [PubMed]
15. Aldeek, F.; McCutcheon, N.; Smith, C.; Miller, J.H.; Danielson, T.L. Dissolution Testing of Nicotine Release from OTDN Pouches: Product Characterization and Product-to-Product Comparison. *Separations* **2021**, *8*, 7. [CrossRef]
16. Lunell, E.; Fagerström, K.; Hughes, J.; Pendrill, R. Pharmacokinetic Comparison of a Novel Non-tobacco-Based Nicotine Pouch (ZYN) With Conventional, Tobacco-Based Swedish Snus and American Moist Snuff. *Nicotine Tob. Res.* **2020**, *22*, 1757–1763. [CrossRef] [PubMed]

17. Lunell, E.; Lunell, M. Steady-state nicotine plasma levels following use of four different types of Swedish snus compared with 2-mg Nicorette chewing gum: A crossover study. *Nicotine Tob. Res.* **2005**, *7*, 397–403. [CrossRef]
18. Caraway, J.W.; Chen, P.X. Assessment of mouth-level exposure to tobacco constituents in U.S. snus consumers. *Nicotine Tob. Res.* **2013**, *15*, 670–677. [CrossRef]
19. Navazesh, M.; Christensen, C.M. A comparison of whole mouth resting and stimulated salivary measurement procedures. *J. Dent. Res.* **1982**, *61*, 1158–1162. [CrossRef]
20. Rahman, Z.; Dharani, S.; Khuroo, T.; Khan, M.A. Potential Application of USP Paddle and Basket Dissolution Methods in Discriminating for Portioned Moist Snuff and Snus Smokeless Tobacco Products. *AAPS PharmSciTech* **2021**, *22*, 51. [CrossRef]
21. Food and Drug Administration. Guidance for Industry: Dissolution Testing of Immediate Release Solid Oral Dosage Forms. Available online: https://www.fda.gov/regulatory-information/search-fda-guidance-documents/dissolution-testing-immediate-release-solid-oral-dosage-forms (accessed on 10 September 2021).
22. ISRCTN44913332. The In-Vivo Extraction of Nicotine and Flavor Compounds from a Single Dose of a Non-Tobacco-Based Nicotine Pouch (ZYN®) Compared with Conventional, Tobacco-Based Swedish Snus among Current, Daily Snus Users. Available online: https://www.isrctn.com/ISRCTN44913332?q=&filters=conditionCategory:Not%20Applicable&page=1&pageSize=10 (accessed on 8 October 2021).
23. Moore, J.W.; Flanner, H.H. Mathematical Comparison of Curves with an Emphasis on in Vitro Dissolution Profiles. *Pharm. Technol.* **1996**, *20*, 64–74.
24. Digard, H.; Errington, G.; Richter, A.; McAdam, K. Patterns and behaviors of snus consumption in Sweden. *Nicotine Tob. Res.* **2009**, *11*, 1175–1181. [CrossRef]
25. Proctor, G.B. The physiology of salivary secretion. *Periodontol 2000* **2016**, *70*, 11–25. [CrossRef]
26. Pijpe, J.; Kalk, W.W.; Bootsma, H.; Spijkervet, F.K.; Kallenberg, C.G.; Vissink, A. Progression of salivary gland dysfunction in patients with Sjogren's syndrome. *Ann. Rheum. Dis.* **2007**, *66*, 107–112. [CrossRef]
27. Digard, H.; Gale, N.; Errington, G.; Peters, N.; McAdam, K. Multi-analyte approach for determining the extraction of tobacco constituents from pouched snus by consumers during use. *Chem. Cent. J.* **2013**, *7*, 55. [CrossRef] [PubMed]

Article

Determination of Nicotine-Related Impurities in Nicotine Pouches and Tobacco-Containing Products by Liquid Chromatography–Tandem Mass Spectrometry

Rozanna Avagyan *, Maya Spasova and Johan Lindholm

Swedish Match AB, Regulatory & Scientific Affairs, Maria Skolgata 83, 118-53 Stockholm, Sweden;
maya.spasova@swedishmatch.com (M.S.); johan.lindholm@swedishmatch.com (J.L.)
* Correspondence: rozanna.avagyan@swedishmatch.com

Abstract: Smokeless tobacco products and nicotine-containing tobacco-free oral pouches have increased in popularity in recent years. They are associated with far fewer health hazards compared to cigarettes. Nicotine pouches are filled with non-tobacco filler and nicotine. The nicotine used in nicotine pouches usually comes from the extraction of tobacco; thus, related alkaloids may be found as impurities at low levels. Moreover, nicotine degradation products are formed because of microbial action, flavor oxidation, exposure to high temperatures etc. Currently, there are no published or recommended methods for the analysis of nicotine degradants in nicotine pouches. Here, we present a sensitive and selective liquid chromatography–tandem mass spectrometry method for the simultaneous determination of seven nicotine-related impurities. All seven analytes and corresponding deuterated internal standards were separated within 3.5 min, including 1 min equilibration. The method was fully validated, showing good linearity with correlation coefficients >0.996 for all analytes, good extraction yields ranging from 78% to 110%, limits of detection between 0.08 and 0.56 µg/g and limits of quantification between 0.27 and 2.04 µg/g. Although the method was mainly developed to determine the degradants of nicotine in nicotine pouches, it was validated and performed well on a broader range of tobacco-containing products.

Keywords: nicotine degradants; nicotine-related impurities; alkaloids; nicotine degradation products; nicotine pouches; reduced-risk products; constituents; method development; method validation

Citation: Avagyan, R.; Spasova, M.; Lindholm, J. Determination of Nicotine-Related Impurities in Nicotine Pouches and Tobacco-Containing Products by Liquid Chromatography–Tandem Mass Spectrometry. *Separations* **2021**, *8*, 77. https://doi.org/10.3390/separations8060077

Academic Editor: Fadi Aldeek

Received: 15 April 2021
Accepted: 28 May 2021
Published: 3 June 2021

Publisher's Note: MDPI stays neutral with regard to jurisdictional claims in published maps and institutional affiliations.

1. Introduction

Nicotine-containing tobacco-free oral pouches belong to a new product category that has gained market shares in recent years [1]. The nicotine pouches are similar to snus, but they contain different non-tobacco fillers and nicotine instead of tobacco leaves. Additionally, the nicotine pouches usually contain pH adjusters, processing aids, artificial sweeteners, flavors, fibers (pouch material) and stabilizers. These products come in a variety of flavors and nicotine content, as well as brand names such as ZYN®, Velo and on!®, manufactured by different manufacturers. Although the long-term health effects of nicotine pouches have not been established yet, it is suggested that they are less harmful than cigarettes [2].

The nicotine used in the manufacturing of nicotine pouches is usually extracted from the tobacco plant; thus, related alkaloids (e.g., nornicotine, anatabine and anabasine) may be found as impurities in small quantities [3]. Moreover, due to environmental factors such as temperature, humidity, light and storage containers, the degradation of nicotine may occur, giving rise to the formation of nicotine degradation products (e.g., cotinine, nicotine-N'-oxide, myosmine and β-nicotyrine) [4]. In the US and European pharmacopoeias, there are recommendations for the purity of nicotine used in pharmaceutical products [5,6] but not in other nicotine-containing products. The nicotine impurities are specified in the European Pharmacopoeia monograph 1452 as nicotine-N'-oxide, cotinine, nornicotine,

31

anatabine, myosmine, anabasine and β-nicotyrine, while the US Pharmacopeia (USP)-grade nicotine requires single impurities to be less than 0.5% (5 mg/g) and total impurities to be less than 1% (10 mg/g) [5,6].

There are several methods to determine the levels of nicotine and its metabolites (e.g., cotinine, nicotine-N'-oxide, nornicotine) using liquid chromatography coupled to tandem mass spectrometry (LC–MS/MS) in human urine [7–9] as well as plasma, semen and sperm by using LC–Orbitrap–MS [10]. Nicotine and related alkaloids (anabasine, anatabine) have also been determined using gas chromatography coupled to flame ionization detection (FID), nitrogen–phosphorus detection (NPD) and MS in tobacco-containing products and tobacco smoke [11–14]. Several methods are also available for the analysis of nicotine-related alkaloids and impurities in electronic cigarette liquids, cartridges and aerosols [4,15–17].

However, there are currently no published or recommended methods available for the analysis of nicotine impurities in nicotine pouches. The above-mentioned methods have not been investigated, and may not be entirely suitable for the analysis of nicotine pouches due to differences in their matrix composition. In this paper, we describe a sensitive and selective method using LC–MS/MS for the simultaneous determination of seven nicotine impurities in four nicotine pouch products, as well as five tobacco products (namely, CORESTA Smokeless Tobacco Reference Products CRP 1.1, CRP 2.1, CRP 3.1, CRP 4.1 and a cigar). Although there are no regulatory requirements or recommendations for these impurities in nicotine products, the method can be used for quality control purposes (e.g., to check the purity of nicotine, as well as for stability studies of nicotine pouches by monitoring the degradation of nicotine).

2. Materials and Methods

2.1. Standards and Reagents

Standards of nicotine-N'-oxide, nornicotine, anabasine, anatabine, cotinine, myosmine, β-nicotyrine, nicotine-N'-oxide-d3, nornicotine-d4, anabasine-d4, anatabine-d4, cotinine-d3, myosmine-d4 and β-nicotyrine-d3 (purity >95% for all standards) were purchased from Toronto Research Chemicals (Toronto, ON, Canada). Acetonitrile (ACN) (HPLC grade), isopropanol (HPLC grade), formic acid (98–100%, p.a. grade), ammonium formate (LC–MS grade), ammonium hydroxide (25%, LC–MS grade) and acetic acid (LC–MS grade) were obtained from VWR, Radnor, PA, USA. Methanol (MeOH) (HPLC grade) was purchased from Fisher Scientific, Waltham, MA, USA. Water was purified using a Milli-Q® Integral 3 (Millipore SAS, Molsheim, France) water purification system equipped with a Millipak® Express 40 0.22 μm membrane filter (Millipore Corp., Burlington, MA, USA).

Stock solutions with concentrations of approximately 1 mg/mL in methanol were prepared for all the standards and the internal standards, respectively. Intermediate standard solutions were prepared from the stock solutions at three concentration levels, 1, 20 and 200 μg/mL. An intermediate standard solution was prepared for the internal standards as well, containing 18.75 μg/mL of β-nicotyrine-d3 and myosmine-d4 and 6.25 μg/mL of residual internal standards. Six (seven for nicotine-N'-oxide) calibration standards dissolved in 0.2% ammonium hydroxide were also prepared. Stock solutions and intermediate standard solutions were stored in a freezer (−18 °C). Calibration standard solutions were stored in a refrigerator (4–6 °C).

2.2. Sample Handling and Preparation

The CRP samples were stored at approximately −20 °C until analyses were performed, as recommended by CORESTA [18]. Prior to analysis, the CRPs were placed in a refrigerator for 24 h and then equilibrated to ambient conditions before opening. After opening, the samples were placed in a sealed container for short-term storage in the refrigerator. The nicotine pouches and cigar were handled in the same way as the CRPs. It was also noticed that the storage of nicotine pouches prior to re-analysis played a significant role in obtaining accurate results. After opening, it is not recommended to store these samples in the freezer.

An amount of 1.0 ± 0.2 g sample was weighed out in a 100 mL Erlenmeyer flask. The nicotine and CRP 1.1 pouches were cut in two lengthwise. CRP 2.1, CRP 3.1 and CRP 4.1 were weighted out as is, without grinding, while the cigar was ground to obtain a homogeneous sample. A total of 100 µL of internal standard solution and 50 mL of extraction solution (100 mM ammonium formate buffer (pH 3)) were added to the sample. The sample was then shaken on an orbital shaker for 40 min at 130 rpm and then allowed to settle for about 5 min to facilitate filtering. A total of 100 µL of sample solution was transferred to a filter vial (0.2 µm Whatman Mini-UniPrep, Fisher Scientific, USA), while 400 µL 0.3 M ammonium hydroxide was added with Multipette.

2.3. Chromatographic and Mass Spectrometric Conditions

The analyses were performed on a UPLC system from Waters Corp., Milford, CT, USA, consisting of an Acquity I-Class UPLC with binary pumps, fitted with an Acquity Sample manager with a cooling system, an auto-injector with a flow-through needle injection and a column switch with a column oven. The chromatographic separation was performed on a Waters Acquity UPLC BEH C_{18} column, 2.1 mm $\times$ 100 mm, 1.7 µm particle size (Part # 186002352) connected to a Waters pre-filter (Assay, Frit, 0.2 µm, 2.1 mm, part. No. 289002078). Mobile phase A was 0.1% ammonium hydroxide, 10 mM ammonium acetate buffer in MQ water; mobile phase B was 0.1% ammonium hydroxide, 10 mM ammonium acetate buffer in ACN. The injection volume was 1 µL, and the mobile phase flow rate was set to 600 µL/min. The gradient condition used was as follows: initial 7% B, 0.2 min 7% B, 1.25 min 45% B, 1.80 min 45% B, 2.20 min 98% B, 2.50 min 98% B and 2.51 min 7% B. The system was equilibrated for 1 min with 7% B before each run.

The MS system was a Waters Xevo TQ-XS, and the MS parameters were set as follows: capillary voltage 0.50 kV, cone voltage 30 V, desolvation 1000 L/h, cone 150 L/h, nebulizer 7 bar, collision gas flow 0.15 mL/min, desolvation temperature 600 °C and source temperature 150 °C. The dwell time for each transition was 0.150 s, except for the transitions of β-nicotyrine and β-nicotyrine-d3 that had dwell times of 0.041 s. Quantitative analyses were performed in MS/MS mode. The analyte-specific parameters are shown in Table 1. Data were acquired and processed with Waters MassLynx (Ver. 4.2.; Waters Corp., Milford, CT, USA).

Table 1. Collision energies, retention times, quantification and confirmation traces.

Compound Name	Collision Energy (eV)	Retention Time (min)	Quantification Trace (m/z)	Confirmation Trace (m/z)
Nicotine-N'-oxide	20 [1]; 14 [2]	~0.45	179.07 > 130.00	179.07 >132.01
Nicotine-N'-oxide-d3	15; 20	~0.45	182.11 > 132.02	182.11 > 130.01
Nornicotine	15; 20	~1.11	149.04 > 129.99	149.04 > 116.99
Nornicotine-d4	10; 20	~1.10	153.13 > 136.05	153.13 > 121.03
Cotinine	22; 20	~1.02	177.04 > 79.95	177.04 > 98.00
Cotinine-d3	22; 20	~1.02	180.08 > 79.96	180.08 > 101.03
Anabasine	18; 16	~1.30	163.06 > 91.97	163.06 > 93.99
Anabasine-d4	22; 22	~1.29	167.12 > 96.02	167.12 > 122.03
Anatabine	12; 12	~1.26	161.05 > 107.00	161.05 > 144.01
Anatabine-d4	15; 15	~1.25	165.11 > 111.04	165.11 > 148.05
Myosmine	20; 20	~1.29	146.99 > 104.99	146.99 > 129.98
Myosmine-d4	30; 22	~1.28	151.08 > 81.97	151.08 > 109.03
β-Nicotyrine	22; 23	~1.68	159.03 > 144.00	159.03 > 117.00
β-Nicotyrine-d3	22; 26	~1.67	162.06 > 144.00	162.06 > 117.06

[1] Collision energy for quantification trace, [2] Collision energy for confirmation trace.

3. Results and Discussion

3.1. UPLC–MS/MS Analysis

The analytes were separated within 2.0 min with a total run time of 3.5 min (including 1 min equilibration) and most peaks were well resolved. Example chromatograms with multiple reaction monitoring (MRM) transitions for the analytes in a standard mixture are shown in Figure 1. Only anabasine and myosmine could not be separated; however, due to different MRM transitions, each of them could be correctly quantified.

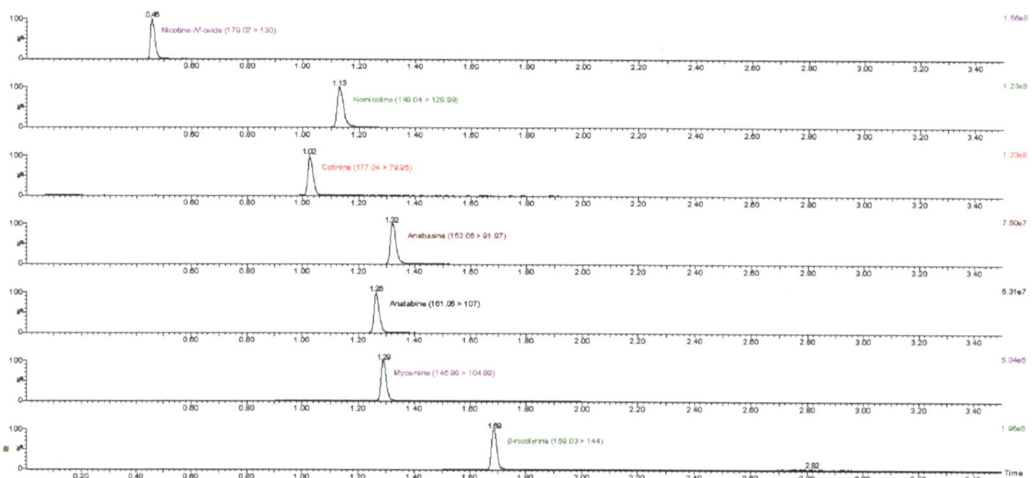

Figure 1. MRM transitions for all analytes in calibration standard 3.

The ionization of the analytes was examined in both positive and negative modes. However, ionization was better in positive mode for all the analytes. The optimal instrumental parameters for each analyte were obtained by tuning, using direct infusion of individual standard solutions. The analytes and the deuterated internal standards were divided into three time windows in order to increase the dwell times and the signal intensity of each compound. Two MRM transitions were generated, for quantification and confirmation purposes, respectively. The identification of the analytes in samples was based on a comparison of MRM transitions and retention times with pure standard solutions. Individual deuterated internal standards were used for each analyte.

3.2. Method Validation

Samples of nine different matrices were used in the method validation procedure, four nicotine pouch products described in Table 2, Swedish-style snus pouches (CRP 1.1), American-style loose moist snuff (CRP 2.1), American-style loose dry snuff powder (CRP 3.1), American-style loose leaf chewing tobacco (CRP 4.1) and a cigar. Since most of the analytes were not detected in the nicotine pouches, the pouches were spiked before the extraction using the intermediate standard solutions. All the matrices were included in the validation experiments to determine repeatability, detection and quantification limits, matrix effects and extraction recoveries. An extended validation was performed using three of the matrices (CRP 1.1, CRP 2.1 and nicotine pouch product 1 (NP1)) to also determine within-laboratory precision and accuracy of the method.

Table 2. Description of nicotine pouch products.

Sample Name	Sample Matrix	Nicotine (mg/g)	Flavor
Nicotine pouch product 1 (NP1)	Granulated filler	4	Spearmint
Nicotine pouch product 2 (NP2)	Plant fibers	12	Spearmint
Nicotine pouch product 3 (NP3)	Fibers from eucalyptus and pine	8	Mint
Nicotine pouch product 4 (NP4)	Plant fiber (cellulose) and chewing gum base	17	Smooth Mint

3.2.1. Linearity and Detection Limits

Linearity was investigated by analyzing six standard solutions three times in a row on the same day in concentrations of 4–800 ng/mL for nornicotine, anatabine and anabasine; in concentrations of 4–400 ng/mL for myosmine, β-nicotyrine and cotinine; and in concentrations of 4–1000 ng/mL for nicotine-N'-oxide. The linearity of all analytes was good with a correlation coefficient >0.996, while the relative residuals were less than 15% when the standard curves were weighted by 1/y.

The limits of detection (LODs) and limits of quantification (LOQs) for the different analytes were calculated in all matrices. The signal/noise (S/N) ratio was measured with RMS (root mean square) in the MassLynx software and was used to calculate the LOD and the LOQ. For the analytes with concentrations <LOD in some matrices, the S/N was calculated from the accuracy data (spiked level 1). The LOD was determined as the concentration where S/N = 3. Similarly, the LOQ was determined as the concentration where S/N = 10. The LOD and the LOQ varied in the different matrices, but Table 3 shows the highest LOD and LOQ values for each analyte.

Table 3. LOD and LOQ for all analytes in μg/g.

Analyte	LOD (μg/g)	LOQ (μg/g)
Nicotine-N'-oxide	0.56	1.86
Nornicotine	0.46	1.53
Cotinine	0.10	0.34
Anatabine	0.18	0.59
Anabasine	0.08	0.27
Myosmine	0.36	1.18
β-Nicotyrine	0.61	2.04

3.2.2. Repeatability, Within-Laboratory Precision and Accuracy

Repeatability was estimated by preparing and analyzing six replicates for each matrix at one time point. The pooled relative standard deviations (%RSDs) are listed in Table 4. For nicotine pouches, due to analyte concentrations <LOD, the estimation of %RSD was based on spiked samples.

Table 4. %RSD pool for repeatability.

Analyte	Repeatability
Nicotine-N'-oxide	4.54
Nornicotine	3.50
Cotinine	5.11
Anabasine	3.65
Anatabine	3.93
Myosmine	10.3
β-Nicotyrine	8.41

Within-laboratory precision was determined by four laboratory technicians analyzing three replicates of three matrices (CRP 1.1, CRP 2.1 and NP1) at six different time points. New extraction solutions, mobile phases and internal standard solutions were also prepared and used. The %RSDs for the different matrices are listed in Table 5 and were higher in NP1 compared to the other two matrices, probably because the same cans were used and re-opened several times during the time the analyses were carried out.

Table 5. Within-laboratory precision (%RSD).

Matrices	Nicotine-N'-oxide	Nornicotine	Anabasine	Anatabine	Myosmine	β-Nicotyrine	Cotinine
CRP 1.1	8.4	5.7	4.3	4.4	11	12	4.5
CRP 2.1	5.1	11	3.6	3.3	5.8	6.0	9.4
NP1	14	18	16	17	20	19	19
%RSD pool	10	11	10	10	14	13	12

Accuracy was determined by spiking three matrices (CRP 1.1, CRP 2.1 and NP1) with all analytes at three concentration levels. Six replicates at each level and six unspiked replicates were analyzed. The analyte concentrations for accuracy experiments were calculated using internal standards and relative response factors. Table 6 provides a summary of the accuracy of all analytes in the three matrices. Cotinine for CRP 2.1 had the lowest accuracy, between 52% and 63%, which is probably due to the matrix composition.

Table 6. Accuracy data (%) for CRP 1.1, CRP 2.1 and NP1.

Analyte	Spiking Levels (µg)	CRP 1.1	CRP 2.1	NP1
Nicotine-N'-oxide	40; 80; 120	101–111	100–108	97–106
Nornicotine	30; 60; 90	84–92	74–81	98–110
Cotinine	10; 20; 30	78–86	52–63	95–107
Anabasine	10; 20; 30	93–97	91–102	103–111
Anatabine	30; 60; 90	90–111	99–110	100–110
Myosmine	5; 15; 30	90–100	91–109	102–106
β-Nicotyrine	10; 20; 30	92–108	89–107	89–103

3.2.3. Matrix Effects and Extraction Yields

In order to investigate the matrix effects and extraction yields, all matrices were spiked with each analyte prior to sample preparation, in prepared extracts and in pure extraction solution.

The absolute matrix effects were determined by comparing the areas of analytes in matrices spiked after sample preparation with areas in standards in pure extraction solution without using the internal standards, by single-point calculation. Unfortified extracts were used for area subtraction for the analytes. The matrix effects are presented in Table 7 and were at reasonable levels for CRP 2.1, CRP 4.1 and the nicotine pouches for most analytes. Ion suppression was observed for CRP 3.1 and cigar matrices for some analytes, while ion enhancement was observed for CRP 1.1 for some analytes. However, the deuterated internal standards compensated well for the matrix effects.

The extraction yields (irrespective of matrix effects in the detector) were determined by comparing the peak areas of analytes in matrices spiked before sample preparation with peak areas in samples spiked after the sample preparation, without using the internal standards, by single point calculation. The extraction yields were good, ranging between 78% and 110% for all analytes and matrices. The extraction yields are presented in Table 8.

Table 7. Matrix effects (%) for all matrices.

Matrices	Nicotine-N'-oxide	Nornicotine	Anabasine	Anatabine	Myosmine	β-Nicotyrine	Cotinine
CRP 1.1	94	148	154	160	130	151	135
CRP 2.1	74	92	112	89	113	118	99
CRP 3.1	28	39	59	40	97	75	26
CRP 4.1	99	37	87	61	97	97	80
Cigar	55	40	77	60	98	98	18
NP1	104	103	104	104	103	105	105
NP2	90	101	103	101	103	106	98
NP3	80	102	107	105	109	110	102
NP4	64	94	106	105	106	104	101

Table 8. Extraction yields (%) for all matrices.

Matrices	Nicotine-N'-oxide	Nornicotine	Anabasine	Anatabine	Myosmine	β-Nicotyrine	Cotinine
CRP 1.1	96	95	93	96	84	81	95
CRP 2.1	100	106	104	105	102	100	102
CRP 3.1	108	110	108	107	96	102	107
CRP 4.1	103	103	106	105	96	78	104
Cigar	104	106	103	102	88	85	102
NP1	93	103	107	97	96	91	104
NP2	103	99	99	100	104	100	100
NP3	99	103	100	10	100	90	97
NP4	98	101	99	98	103	92	96

3.2.4. Cross Talk/Carry-Over

To verify that cross talk did not occur, the analyte solutions were injected without internal standards, and the MRM transitions for the analytes and internal standards were monitored to verify that no ions were detected from the analytes giving rise to a peak for the internal standards and the reverse. No (negligible, <1% of the standard peak) peaks were detected; consequently, it can be concluded that cross talk did not occur.

Carry-over was checked by injecting the strongest calibration standard. A blank was injected after the calibration standard. No (negligible, <1% of the standard peak) peaks were detected in the blank injections, which is consistent with no carry-over effect.

3.2.5. Stability of Sample Extracts and Standard Solutions

The stability of the prepared samples in the auto-injector (4 °C) or refrigerator (4–6 °C) was investigated by analyzing samples immediately after preparation and after 3, 7 and 14 days. The samples were stored in vials with perforated and unperforated septa. The results revealed that the samples were stable for at least seven days in vials with unperforated septa and only three days in vials with perforated septa.

The stability of stock and calibration standard solutions was investigated as well, showing a shelf life of 1 year for stock and intermediate solutions stored in a freezer (−18 °C) and 6 months for the calibration standards stored in a refrigerator (4–6 °C).

3.3. Analysis of Samples

As mentioned above, nine different matrices were used in the method validation procedure. All the analytes were detected in the tobacco-containing matrices (CRP 1.1–CRP 4.1 and cigar). Most of the analytes were also detected in the nicotine pouch products, however, they were at lower levels compared to tobacco-containing matrices. Nicotine-N'-oxide was detected in all nicotine pouch products, while β-nicotyrine was not detected in any of the nicotine pouch products. Almost all the analytes (except for β-nicotyrine) were detected in NP4, while only nicotine-N'-oxide was detected in NP1. The analyte concentrations of nicotine-N'-oxide ranged from 2.6 µg/g to 820 µg/g, for nornicotine from 2.1 to 340 µg/g, for anatabine from 1.2 to 260 µg/g, for cotinine from 1.2 to 130 µg/g, for β-nicotyrine from 3.3 to 59 μg/g, for anabasine from 2.9 to 56 μg/g and for myosmine from 1.3 to 10 μg/g in the different matrices. Table 9 shows a summary of all analytes and their concentrations in the nine matrices. A representative chromatogram of a sample, CRP 1.1, is shown in Figure 2. The highest analyte concentrations were detected in the cigar and CRP 3.1 matrices, except for nornicotine, with the highest concentration detected in CRP 4.1. Except for nicotine pouches, CRP 1.1 had the lowest concentrations of all analytes.

Table 9. Determined concentrations (μg/g) of the analytes (n = 3) in the different matrices.

Matrices	Nicotine-N'-oxide	Nornicotine	Anabasine	Anatabine	Myosmine	β-Nicotyrine	Cotinine
CRP 1.1	120[1] (6.6)[2]	96 (4.0)	22 (1.0)	100 (3.5)	3.9 (0.7)	3.3 (0.2)	24 (1.6)
CRP 2.1	602 (2.9)	110 (3.3)	45 (1.6)	210 (10)	4.8 (0.4)	20 (1.0)	31 (2.0)
CRP 3.1	800 (3.8)	180 (6.8)	56 (1.5)	260 (4.4)	5.5 (0.3)	59 (2.1)	62 (3.0)
CRP 4.1	460 (13)	340 (10)	39 (1.7)	250 (11)	6.1 (0.6)	14 (1.8)	45 (1.7)
Cigar	820 (6.3)	250 (8.3)	42 (2.0)	180 (8.2)	10 (0.9)	14 (1.4)	130 (5.5)
NP1	2.6 (0.1)	n.d.[3]	n.d.	n.d.	n.d.	n.d.	n.d.
NP2	57 (1.2)	2.1 (0.04)	n.d.	n.d.	n.d.	n.d.	n.d.
NP3	74 (1.8)	5.0 (0.08)	n.d.	1.4 (0.04)	1.3 (0.11)	n.d.	n.d.
NP4	62 (1.1)	9.8 (0.06)	2.9 (0.1)	1.2 (0.07)	4.3 (0.09)	n.d.	1.2 (0.04)

[1] Concentration; [2] standard deviation; [3] not detected.

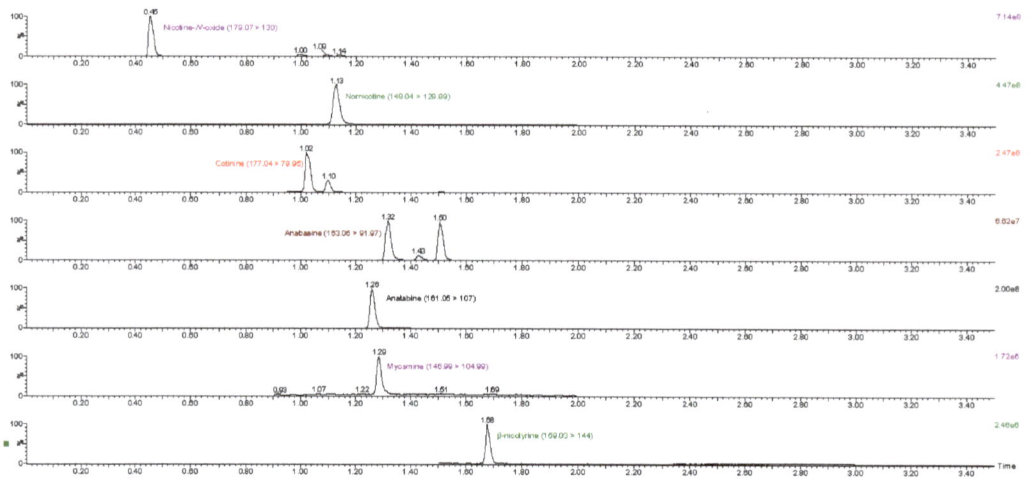

Figure 2. MRM transitions of analytes detected in CRP 1.1.

4. Conclusions

A simple and rapid method for the analysis of nicotine-related impurities using UPLC–MS/MS was developed in the present study for nicotine pouch products and five other tobacco-containing matrices. The simultaneous determination of seven nicotine impurities and seven internal standards with a total run time of 3.5 min could be performed with high precision and low LOD and LOQ. Extraction recoveries were good, and matrix effects were small for most of the matrices used in the validation. Although the method was mainly developed to determine nicotine impurities in nicotine pouches, it was validated and performed well for a broader range of nicotine-containing matrices. All the analytes were detected in varying concentrations in the different matrices; however, the concentrations of analytes were lower in the nicotine pouch products compared to the tobacco-containing matrices. There are several methods available for the determination of nicotine degradants, metabolites and alkaloids in various matrices, but this method was developed and adjusted for the analysis of nicotine pouches and the relatively low concentrations of analytes that might be present there [5,6]. Another advantage of this method is that the corresponding deuterated internal standards were used for all the analytes, which compensate well for both the losses in the extraction procedure and the matrix effects. This method could be useful for quality control purposes (e.g., to check the purity of nicotine), as well as for stability studies of nicotine pouches by monitoring nicotine degradation. The method could also be used to compare nicotine pouches with tobacco-containing products (e.g., CRPs).

Author Contributions: Conceptualization, R.A. and J.L.; methodology, R.A. and M.S.; formal analysis, R.A.; validation, R.A.; writing—original draft preparation, R.A.; writing—review and editing, R.A., M.S. and J.L.; visualization, M.S. All authors have read and agreed to the published version of the manuscript.

Funding: This research received no external funding.

Institutional Review Board Statement: Not applicable.

Informed Consent Statement: Not applicable.

Data Availability Statement: Data can be provided by authors upon request.

Acknowledgments: Johan Patring, Johan Redeby and Anna Masser are acknowledged for kindly reading and reviewing the article.

Conflicts of Interest: The authors declare no conflict of interest.

References

1. Delnevo, C.D.; Hrywna, M.; Miller Lo, E.J.; Wackowski, O.A. Examining market trends in smokeless tobacco sales in the United States: 2011–2019. *Nicotine Tob. Res.* **2020**. [CrossRef] [PubMed]
2. Fisher, M.T.; Tan-Torres, S.M.; Gaworski, C.L.; Black, R.A.; Sarkar, M.A. Smokeless tobacco mortality risks: An analysis of two contemporary nationally representative longitudinal mortality studies. *Harm. Reduct. J.* **2019**, *16*, 27. [CrossRef] [PubMed]
3. Hukkanen, J.; Jacob, P.; Benowitz, N.L. Metabolism and Disposition Kinetics of Nicotine. *Pharmacol. Rev.* **2005**, *57*, 79–115. [CrossRef] [PubMed]
4. Etter, J.F.; Zäther, E.; Svensson, S. Analysis of refill liquids for electronic cigarettes. *Addiction* **2013**, *108*, 1671–1679. [CrossRef]
5. United States Pharmacopeia and the National Formulary. *USP 38-NF 33*; The United States Pharmacopeial Convention Inc.: Rockville, MD, USA, 2015.
6. Council of Europe. *European Pharmacopoeia 7.0*; European Directorate for the Quality of Medicines and Healthcare: Strasbourg, France, 2012.
7. Marclay, F.; Saugy, M. Determination of nicotine and nicotine metabolites in urine by hydrophilic interaction chromatography–tandem mass spectrometry: Potential use of smokeless tobacco products by ice hockey players. *J. Chromatogr. A* **2010**, *1217*, 7528–7538. [CrossRef]
8. Xu, X.; Iba, M.M.; Weisel, C.P. Simultaneous and sensitive measurement of anabasine, nicotine, and nicotine metabolites in human urine by liquid chromatography–tandem mass spectrometry. *Clin. Chem.* **2004**, *50*, 2323–2330. [CrossRef] [PubMed]
9. Meger, M.; Meger-Kossien, I.; Schuler-Metz, A.; Janket, D.; Scherer, G. Simultaneous determination of nicotine and eight nicotine metabolites in urine of smokers using liquid chromatography–tandem mass spectrometry. *J. Chromatogr. B* **2002**, *778*, 251–261. [CrossRef]
10. Abu-awwad, A.; Arafat, T.; Schmitz, O.J. Simultaneous determination of nicotine, cotinine, and nicotine N-oxide in human plasma, semen, and sperm by LC-Orbitrap MS. *Anal. Bioanal. Chem.* **2016**, *408*, 6473–6481. [CrossRef] [PubMed]
11. Sheng, L.Q.; Ding, L.; Tong, H.W.; Yong, G.P.; Zhou, X.Z.; Liu, S.M. Determination of nicotine-related alkaloids in tobacco and cigarette smoke by GC-FID. *Chromatographia* **2005**, *62*, 63–68. [CrossRef]
12. Cai, J.; Liu, B.; Lin, P.; Su, Q. Fast Analysis of Nicotine Related Alkaloids in Tobacco and Cigarette Smoke by Megabore Capillary Gas Chromatography. *J. Chromatogr. A* **2003**, *1017*, 187–193. [CrossRef] [PubMed]
13. Yang, S.S.; Smetena, S.I.; Huang, C. Determination of Tobacco Alkaloids by Gas Chromatography with Nitrogen–Phosphorus Detection. *Anal. Bioanal. Chem.* **2002**, *373*, 839–843. [CrossRef] [PubMed]
14. Lisko, J.G.; Stanfill, S.B.; Duncan, B.W.; Watson, C.H. Application of GC-MS/MS for the Analysis of Tobacco Alkaloids in Cigarette Filler and Various Tobacco Species. *Anal. Chem.* **2013**, *85*, 3380–3384. [CrossRef] [PubMed]
15. Liu, X.; Joza, P.; Rickert, B. Analysis of nicotine and nicotine-related compounds in electronic cigarette liquids and aerosols by liquid chromatography-tandem mass spectrometry. *Beiträge Tab. Int. Contrib. Tob. Res.* **2017**, *27*, 154–167. [CrossRef]
16. Famele, M.; Palmisani, J.; Ferranti, C.; Abenavoli, C.; Palleschi, L.; Mancinelli, R.; Fidente, R.M.; de Gennaro, G.; Draisci, R. Liquid chromatography with tandem mass spectrometry method for the determination of nicotine and minor tobacco alkaloids in electronic cigarette refill liquids and second-hand generated aerosol. *J. Sep. Sci.* **2017**, *40*, 1049–1056. [CrossRef] [PubMed]

17. Flora, J.W.; Wilkinson, C.T.; Sink, K.M.; McKinney, D.L.; Miller, J.H. Nicotine-related impurities in e-cigarette cartridges and refill e-liquids. *J. Liq. Chrom. Relat. Tech.* **2016**, *39*, 821–829. [CrossRef]
18. CORESTA. Available online: https://www.coresta.org/coresta-smokeless-tobacco-reference-products (accessed on 6 April 2021).

separations

MDPI

Article

Market Survey of Modern Oral Nicotine Products: Determination of Select HPHCs and Comparison to Traditional Smokeless Tobacco Products

Joseph J. Jablonski, Andrew G. Cheetham and Alexandra M. Martin *

Enthalpy Analytical, LLC, 1470 E. Parham Rd, Richmond, VA 23228, USA; joseph.jablonski@enthalpy.com (J.J.J.); andrew.cheetham@enthalpy.com (A.G.C.)
* Correspondence: alex.martin@enthalpy.com

Abstract: In an effort to combat the risks associated with traditional tobacco products, tobacco product innovation has been redirected towards reducing the consumer's potential exposure to harmful or potentially harmful constituents (HPHCs). Among these innovations are modern oral nicotine products (MONPs). This product class aims to deliver nicotine while limiting the consumer's potential toxicant exposure. This body of work sought to investigate the potential for select HPHC exposure (tobacco-specific nitrosamines, carbonyls, benzo[*a*]pyrene, nitrite, and metals) from MONPs and to compare it to that from traditional tobacco products. This work expands on previously published studies both in terms of diversity of products assessed and analytes tested. In total, twenty-one unique MONPs were assessed and compared to four traditional tobacco products. We found that there was a difference in the potential exposure based on the MONP filler—plant material vs. granulate/powder. Typically, the HPHC levels observed in plant-based MONPs were higher than those observed for granulate/powder products, most notably within the metals analysis, for which the levels were occasionally greater than those seen in traditional smokeless tobacco products. Generally, the overall HPHC levels observed in MONP were at or below those levels observed in traditional tobacco products.

Keywords: modern oral nicotine products; HPHCs; reduced-risk products; product characterizations

Citation: Jablonski, J.J.; Cheetham, A.G.; Martin, A.M. Market Survey of Modern Oral Nicotine Products: Determination of Select HPHCs and Comparison to Traditional Smokeless Tobacco Products. *Separations* **2022**, *9*, 65. https://doi.org/10.3390/separations9030065

Academic Editor: Josef Cvačka

Received: 27 January 2022
Accepted: 25 February 2022
Published: 2 March 2022

Publisher's Note: MDPI stays neutral with regard to jurisdictional claims in published maps and institutional affiliations.

1. Introduction

It is generally acknowledged that use of tobacco products is associated with risks. In an effort to combat these risks, tobacco science and production have refocused their efforts to provide consumers with products that may limit their potential exposure to harmful or potentially harmful constituents (HPHCs). In 2009, the Family Smoking Prevention and Tobacco Control Act ('Tobacco Control Act') was passed in which control over regulatory oversight was given to the US Food and Drug Administration (FDA) [1,2]. Included in this act were specific requirements for the language to be included on warning labels for various tobacco products and the need for scientific rigor when making claims for any modified risk profile a product may offer [3]. To be able to claim a modified risk profile, manufacturers must submit scientific evidence to support the claim as part of a Modified-Risk Tobacco Product (MRTP) application, and FDA permission must be received. The Tobacco Control Act further required the FDA to establish a list of harmful or potentially harmful constituents to human health found in mainstream smoke and tobacco products (referred to as the 'HPHC list') [4].

Reducing the consumer's exposure to compounds on the HPHC list is one way that risk can conceivably be lowered. Given that combustion is the main source for many of the HPHC compounds, alternative means of delivering nicotine are being promoted and developed. Examples of such products are electronic nicotine delivery systems (ENDS) and heated tobacco products (HTP), which both produce aerosolized nicotine for inhalation.

An alternative to smoking altogether is the use of traditional smokeless tobacco products (STPs), such as chewing tobacco and snuff. STPs have seen a 23.1% increase in total usage over the same period that cigarette consumption has declined by 38.7%, though whether the two are linked is unclear given that there is a perception among some US smokers that STPs do not provide any reduction in toxicant exposure [5–7]. Indeed, STPs are known to contain a number of HPHCs categorized by the IARC (International Agency for Research on Cancer) as being Group 1 carcinogens (e.g., formaldehyde, *N*-Nitrosonornicotine (NNN), 4-(*N*-Methylnitroamino)-1-(3-pyridyl)-1-butanone (NNK), and cadmium) [8–11]. The production techniques for STPs vary considerably and can include curing, fermentation, and pasteurization processes that will influence the HPHC profile of the resulting product, creating a broad spectrum of potential risk among this product class. For instance, typical US moist snuff (including snus) uses a fermentation process that leads to high levels of tobacco-specific nitrosamines (TSNAs) whereas Swedish-style moist snuff is heat-treated (pasteurized) and contains lower TSNA levels [12]. Of the few tobacco products that have been granted MRTP designation by the FDA, eight are Swedish-style snus products manufactured by Swedish Match USA, Inc., which can state they present a reduced risk for certain cancers and diseases when compared with cigarettes [13,14]. Similar to cigarettes, there has been an emergence of alternative products that replicate the STP usage experience with a lowered HPHC exposure risk, namely modern oral nicotine products.

Modern oral nicotine products (MONPs), also known as tobacco-free nicotine products (TFNPs), are a novel class of nicotine-containing products aimed at further reducing toxicant exposure while still delivering the desired nicotine dosage. These products are intended to be consumed in a similar way as STPs, with placement between the gum and the lip/cheek. MONPs are produced in two main formulations—white granular powders (WGP) and plant-based versions. The white granular powder MONPs are pre-portioned and composed of a number of ingredients that include a stabilized form of nicotine (e.g., nicotine salt, nicotine-polacrilex), pH-adjusting agents (e.g., sodium carbonates), filler materials (e.g., modified cellulose, microcrystalline cellulose), sweeteners, and flavorings. Plant-based MONPs more closely mimic traditional STPs in that they are moist products produced using many of the same techniques and are packaged as either long-cut (loose) or pre-portioned pouches. Plant-based MONPs differ from STPs in that they are made from non-tobacco plant-based materials with pharmaceutical-grade nicotine added during the production process. Both forms of MONPs are typically sold in a variety of flavors with some also having the option of multiple nicotine strengths, granting the consumer a variety of options to choose from. At the time of publication, no MRTP applications for any MONPs have been made public on the FDA website (as is required by the Tobacco Control Act).

With the requirements set forth by the FDA pertaining to the classification of modified-risk products, the challenges of accurately assessing potential HPHC exposure comes to the forefront of any analytical testing laboratory. Currently, there are a number of published, standardized methods for the analysis of HPHCs found in smokeless tobacco products. However, as potential HPHC levels are expected to be low, the question of whether these standardized methods are "fit for purpose" must be addressed. A 2021 collaborative study undertaken by the Tobacco and Tobacco Products Analytes (TTPA) sub-group of CORESTA (Cooperation Centre for Scientific Research Relative to Tobacco) examined the suitability of existing CORESTA recommended methods (CRMs) for the analysis of select HPHCs (nicotine, TSNAs, carbonyls, benzo[*a*]pyrene (B[*a*]P), and metals (arsenic and cadmium)) in nicotine pouches [15]. This study found that the methods were suitable and the nicotine pouch matrix was subsequently added to the scope of the respective CRMs in December 2021.

A survey of the current literature turns up few studies that focus on the assessment of MONPs and their potential HPHC exposure risk [16–20]. Further, these studies tend to have a narrow focus on a single product brand, limiting their overall scope. It is the intention of this publication to address this deficiency. Herein we describe the screening of seven brands of modern oral nicotine products. Within each brand, where possible, multiple flavors were analyzed. In addition to these products, two CORESTA smokeless tobacco reference products were screened, along with two smokeless tobacco products that are currently on the US market. In all, 25 unique products were assessed for a select list of HPHCs including TSNAs, carbonyls, nitrite, benzo[*a*]pyrene, and metals.

2. Materials and Methods

2.1. Materials

Standards were prepared either from neat materials (Sigma-Aldrich (St. Louis, MO, USA), Alfa Aesar (Tewksbury, MA, USA)) or an ISO 17034-certified reference standard solution containing the analytes of interest (Cerilliant (Round Rock, TX, USA), Toronto Research Chemicals (North York, ON, Canada), Inorganic Ventures (Christiansburg, VA, USA), Restek (Bellefonte, PA, USA), Accustandard (New Haven, CT, USA)). Where required, labelled internal standards were obtained from CDN Isotopes (Pointe-Clare, QC, Canada). For sample preparation, all reagents were sourced through Thomas Scientific (Swedesboro, NJ, USA) and were of American Chemical Society (ACS) grade or better where available, except for Type 1 water (18.2 MΩ·cm), which was generated in-house [21].

2.2. Test Products

All modern oral nicotine products were purchased by the authors through the online retailers, Northerner and Nicokick, with the exception of Black Buffalo products, which were purchased directly from the manufacturer. All test products were stored refrigerated and brought to room temperature prior to extraction.

2.3. Method Summaries

All product analysis was conducted using Enthalpy's in-house methods, which are fully validated for the analysis of smokeless tobacco products and have been shown to be suitable for the analysis of modern oral nicotine products. Where the pouch weight exceeded the stated sample size, a single pouch was extracted and analyzed. All pouched products were cut in half prior to extraction, with both the filler and pouch material analyzed.

2.3.1. Nicotine Analysis

Nicotine was assessed for products where levels were not reported on the packaging from the manufacturer. The method used for analysis of nicotine was based upon the Centers for Disease Control and Prevention (CDC) method [22] and the CORESTA Recommended Method No. 62 (CRM 62) [23]. In brief, an aliquot of 2N sodium hydroxide was added to a pre-weighed sample (approximately 1.0 g). Methyl-*t*-butyl ether containing quinoline was added and nicotine was extracted into the organic layer via solvent–solvent extraction with mechanical shaking. An aliquot of the organic layer was transferred to a sample vial for analysis via GC-FID.

The analysis of nicotine for select products in this study was carried out using an Agilent (Santa Clara, CA, USA) 6890 gas chromatograph equipped with a flame ionization detector (FID). The analytical column used for analysis was a HP-5, 30 m × 0.32 mm ID × 0.25 μm, with a carrier gas (helium) flow rate of 1.7 mL/min. The injection volume was 2 μL, split 40:1 with an injection port temperature of 250 °C. The following GC oven temperature program was used: initial oven temperature of 110 °C, no hold; 10 °C/min to 185 °C, no hold; ramp 6 °C/min to 240 °C, hold for 10 min. The detector temperature was set to 250 °C.

The calibration curve was constructed with seven points using a linear calibration model with 1/x weighting. The calibration range was 0.0356 to 1.19 mg/mL.

2.3.2. Benzo[*a*]pyrene Analysis

Benzo[*a*]pyrene (B[*a*]P) was extracted from approximately 1.0 g of tobacco or MONP with methanol and mechanical shaking. After centrifuging the sample, the supernatant was collected and evaporated to approximately 1 mL. The sample was then filtered prior to analysis via UPLC-FLR (Ultra Performance Liquid Chromatography with Fluorescence detection). B[*a*]P was quantitated using B[*a*]P-d_{12} as the internal standard.

UPLC was carried out using a Waters (Milford, MA, USA) H-class quaternary pump with a flowthrough needle sample manager. The analytical column used for analysis was an Agilent Zorbax RRHD Eclipse PAH column (150 mm × 2.1 mm, 1.8 μm) with an Agilent Zorbax Eclipse PAH guard column (2.1 mm × 5 mm, 1.8 μm). A 10 μL injection volume was used with a flow rate of 0.5 mL/min and a column temperature of 45 °C. The injection run time was 10 min with an isocratic gradient profile consisting of 20% water and 80% acetonitrile. Samples were detected using a Waters ACQUITY fluorescence detector with an excitation wavelength of 364 nm and an emission wavelength of 405 nm.

The calibration curve was constructed with ten points using a linear calibration model with 1/x weighting. The calibration range used was 0.10 to 100.4 ng/mL.

2.3.3. Nitrite Analysis

Nitrite was extracted from approximately 2 g of tobacco or MONP using water and mechanical shaking. The extracts were centrifuged and the supernatant was filtered prior to analysis via a continuous flow analyzer (CFA). During analysis, nitrite permeated through a dialysis membrane and reacted with sulfanilamide to form a diazonium ion, which was further coupled with *N*-1-naphthylethyldiamine dihydrogen chloride (NED) to form a purple azo dye. Absorbance was measured at 540 nm and sample extracts were quantitated using external calibration.

The analysis of tobacco extracts was performed using an Astoria 2 continuous flow analyzer (Astoria-Pacific, Clackamas, OR, USA) with a nitrite manifold and 540 nm filter.

The calibration curve was constructed with six points using a first-order polynomial calibration model with no weighting. The calibration range used was 0.10 to 3.0 μg/mL.

2.3.4. Tobacco-Specific Nitrosamine (TSNA) Analysis

The method of extraction was based upon the CORESTA Recommended Method No. 72 (CRM 72) [24]. TSNAs were extracted from approximately 1 g of tobacco or MONP using an aqueous solution of ammonium acetate and shaken mechanically. Extracts were subsequently filtered prior to analysis by UPLC-MSMS. The level of TSNAs present in each brand/sample was quantified using deuterated analogs of each analyte as internal standards (NAB-d_4, NAT-d_4, NNK-d_4, and NNN-d_4).

UPLC analysis was carried out using a Waters ACQUITY binary pump. The analytical column used was a Waters ACQUITY BEH C18 (2.2 mm × 100 mm, 1.7 μm) column with matching pre-column. The mobile phases used were A: 0.01% acetic acid in water and B: 0.1% acetic acid in methanol. An injection volume of 10 μL was used with a flow rate of 0.2 mL/min. The gradient profile was: 0–1 min 99% solvent A; 1–4 min 99–10% solvent A; 4–4.01 min 10–1% solvent A; 4.01–5.75 min 1% solvent A; 5.75–5.9 min 1–99% solvent A; 5.9–8 min 99% solvent A. A constant column temperature of 55 °C was maintained throughout the injection.

Detection of the TSNAs was carried out using a Waters (Milford, MA, USA) Xevo-TQ detector. Samples were analyzed using electrospray ionization operating in positive ion mode (ESI+). The source temperature was 150 °C and the desolvation temperature was 500 °C. Nitrogen was used as the desolvation gas (1000 L/h) and argon was used as the collision gas. Analytes were detected in multiple reaction monitoring mode with specific analysis parameters described in Table 1.

Table 1. MS/MS instrument parameters for the detection of TSNAs.

Analyte	Precursor Ion (*m/z*)	Product Ion (*m/z*)	Cone Voltage (V)	Collision Energy (V)
NAB [1]	192.9	162.1	16	10
NAT [1]	190.2	160.1	14	15
NNK [1]	208.2	122.0	16	10
NNN [1]	178.2	148.1	16	14
NAB-d$_4$	196.2	166.1	12	15
NAT-d$_4$	194.2	164.0	13	14
NNK-d$_4$	212.2	126.2	18	12
NNN-d$_4$	182.3	152.1	16	10

[1] Tobacco-specific nitrosamines (TSNAs): NNK = 4-(methylnitrosamino)-1-(3-pyridyl)-1-butanone; NNN = *N'*-nitrosonornicotine; NAB = *N'*-nitrosoanabasine; NAT = *N'*-nitrosoanatabine.

The calibration curve was constructed with eight points using a quadratic calibration model with 1/x weighting. The calibration range for each of the TSNAs was as follows: NAT, NNK, NNN: 0.48 to 300 ng/mL; NAB: 0.12 to 75 ng/mL.

2.3.5. Carbonyls Analysis

The method of extraction was based upon the CORESTA Recommended Method No. 86 (CRM 86) [25]. Approximately 1 g of tobacco or MONP was suspended in acidic aqueous ammonium formate buffer solution. Any carbonyls that were present were derivatized using 2,4-dinitrophenylhydrazine (DNPH) and then extracted into hexanes in situ via solvent–solvent extraction. The hexane layer was transferred to an autosampler vial for analysis via UPLC-MS/MS. Quantitation was performed using the deuterated internal standards formaldehyde-d$_2$, acetaldehyde-d$_4$, and crotonaldehyde-DNPH-d$_3$.

UPLC was carried out using a Waters ACQUITY binary pump. The analytical column used for analysis was a Waters ACQUITY BEH Shield RP18 column (2.1 mm × 100 mm, 1.7 μm) with a Waters BEH C18 guard column (2.1 mm × 5 mm, 1.7 μm). The mobile phases used were A: 1 mM acetic acid in water and B: 1 mM acetic acid in 93:7 (*v/v*) methanol:acetonitrile. A 2 μL injection volume was used with a constant flow rate of 0.35 mL/min and a column temperature of 50 °C. The gradient profile used was: 0.0–4.0 min 55–40% solvent A; 4.0–5.5 min 40–37% solvent A; 5.5–7.25 min 37–25% solvent A; 7.25–7.27 min 25–0% solvent A; 7.27–8.25 min hold at 0% solvent A; 8.25–8.27 min 0–55% solvent A; 8.27–10.25 min 55% solvent A.

Detection of the carbonyls was carried out using a Waters Xevo-TQ mass spectrometer. Samples were analyzed using electrospray ionization operating in negative ion mode (ESI-). Source parameters included a 2.00 kV capillary voltage, 150 °C source temperature, and a desolvation temperature of 500 °C. Nitrogen was used as the desolvation gas (1000 L/h) and argon was used as the collision gas. Analytes were detected in multiple reaction monitoring mode and compound specific parameters can be found in Table 2.

Table 2. MS/MS instrument parameters for the detection of cabonyls.

Analyte	Precursor Ion (*m/z*)	Product Ion (*m/z*)	Cone Voltage (V)	Collision Energy (V)
Acetaldehyde	223.1	151.1	18	10
Crotonaldehyde	249.05	172.1	20	14
Formaldehyde	209.05	163.1	14	6
Acetaldehyde-d$_4$	227.15	151.1	18	8
Crotonaldehyde-DNPH-d$_3$	252.05	175.2	18	14
Formaldehyde-d$_2$	210.95	163.1	16	8

The calibration curve was constructed with eight points using a linear calibration model with 1/x weighting. The calibration range for each analyte was as follows: acetaldehyde: 0.0101 to 2.01 μg/mL; crotonaldehyde: 0.00548 to 0.199 μg/mL; formaldehyde: 0.0101 to 2.01 μg/mL.

2.3.6. Metals Analysis

The method of extraction was a modified version of the CORESTA Recommended Method No. 93 (CRM 93) [26]. Metals were quantitated from an aqueous digestion of approximately 0.5 g of tobacco or MONP. The digestion was performed using concentrated trace-metal-grade nitric acid utilizing a microwave followed by centrifugation. Samples were quantified by ICP-MS equipped with a dynamic reaction cell. Internal standards were used to correct for instrumental drift and were tailored to the metal being quantitated.

Metals analysis was performed on an Agilent (Santa Clara, CA, USA) 7700 ICP-MS. Self-tuning was performed each day of analysis and 72-Ge, 103-Rh, and 209-Bi were used as internal standards. A complete list of instrument parameters can be found in Table 3 and the masses of measure can be found in Table 4.

Table 3. ICP-MS operating parameters for the detection of metals.

Parameter	Mode		
Gas Mode	No Gas	He	H_2
RF Power		1600 W	
RF Matching		1.80 V $\pm$ 0.20	
Carrier Gas Flow		0.75 L/min	
Dilution Gas Flow		0.25 L/min	
S/C Temperature		2 °C	
ORS Gas	NA	Helium	Hydrogen
ORS Gas Flow Rates	NA	4.3 mL/min	6.0 mL/min + (0.5 mL/min He)

Table 4. ICP-MS mass/elements of analytes measured in each operating mode.

Analyte	Mass	Gas Mode	Integrations/Mass (s)	Internal Standard (Mass-Element)
Be	9	No Gas	0.51	72-Ge
Cr	52	He	0.51	72-Ge
Co	59	He	0.51	72-Ge
Ni	60	He	0.51	72-Ge
As	75	He	0.51	72-Ge
Se	78	H_2	0.51	72-Ge
Cd	111	He	0.51	103-Rh
Pb	208	He	0.51	209-Bi

The calibration range for the metal analytes was as follows: As, Ni: 0.10 to 50 ng/mL; Be, Cd, Cr, Co: 0.05 to 50 ng/mL; Pb: 0.05 to 30 ng/mL; Se: 0.20 to 50 ng/mL.

3. Results and Discussion

3.1. Study Design

Current product innovation is focused on reducing consumer exposure to HPHCs through the development of modified-risk products. As these products aim to reduce the consumer's exposure to harmful or potentially harmful constituents, the expected levels of each analyte examined is expected to be lower than those observed in traditional smokeless tobacco products. As MONPs are a relatively new product class, most of the literature published are by the product manufacturers and only focus on their own products, which can make cross product comparisons difficult. This study sought to address this by providing a direct comparison within a single study.

For this work, twenty-five (25) different products from nine individual manufacturers were selected for evaluation (Table 5). Of these twenty-five, four were traditional smokeless tobacco products chosen for comparative purposes. The goal was to obtain products from a range of manufacturers that also varied in flavor and product type (pouch vs. long cut). An effort was made to diversify the flavor profiles, but also to ensure flavor overlap between manufacturers when possible. It must be noted that given the variety of nicotine strengths, not only from brand to brand, but even within a manufacturer's own flavor, the products selected for analysis were at the maximum nicotine strength available. As this is a relatively new product classification, consideration of overall market share was not a top priority in the selection of brands. As mentioned previously, MONP are produced in two main forms: white granular powders (pouches) and plant-based (pouched or long cut). Based on these forms of consumption (pouch/long cut), four traditional smokeless tobacco products, two commercial brands and two CORESTA reference products (CRP1.1 and CRP2.1), were selected as the comparators. The commercial pouch product, General Wintergreen White Portion, was chosen as this product has received permission from the FDA to be marketed as a modified-risk tobacco product [13]. Each of these products was screened for several HPHCs that are shown in Table 6, along with their IARC designations.

Table 5. Modern oral nicotine products and smokeless tobacco comparators tested in this study.

Product Name	ID	Product Basis [1]	Portion Weight (g)	Nicotine Content (mg/portion)	Manufacturer
Velo Max Nicotine Pouches, 7 mg Wintergreen	A1	WGP	0.38 [2]	7 [2]	R.J. Reynolds Vapor Co. [5]
Velo Max Nicotine Pouches, 7 mg Black Cherry	A2	WGP	0.38 [2]	7 [2]	R.J. Reynolds Vapor Co. [5]
Velo Nicotine Pouches, 4 mg Mint	A3	WGP	0.44 [2]	4 [2]	R.J. Reynolds Vapor Co. [5]
Velo Nicotine Pouches, 4 mg Citrus	A4	WGP	0.44 [2]	4 [2]	R.J. Reynolds Vapor Co. [5]
on! Nicotine Pouches, 8 mg Wintergreen	B1	WGP	0.26 [2]	8 [2]	Helix Innovations, LLC [6]
on! Nicotine Pouches, 8 mg Citrus	B2	WGP	0.26 [2]	8 [2]	Helix Innovations, LLC [6]
on! Nicotine Pouches, 8 mg Berry	B3	WGP	0.26 [2]	8 [2]	Helix Innovations, LLC [6]
Rogue Pouches, 6 mg Mango	C1	WGP	0.68 [2]	6 [2]	Rogue Holdings, LLC [7]
Rogue Pouches, 6 mg Honey Lemon	C2	WGP	0.68 [2]	6 [2]	Rogue Holdings, LLC [7]
Rogue Pouches, 6 mg Wintergreen	C3	WGP	0.68 [2]	6 [2]	Rogue Holdings, LLC [7]
FRÉ, 12 mg Lush	D1	WGP	0.50 [2]	12 [2]	Nu-X Ventures, LLC [8]
Zyn Nicotine Pouches, 6 mg Citrus	E1	WGP	0.40 [2]	6 [2]	Swedish Match [9]
Zyn Nicotine Pouches, 6 mg Wintergreen	E2	WGP	0.40 [2]	6 [2]	Swedish Match [9]
Fully Loaded, Wintergreen Pouches	F1	P	1.54 [3]	8.0 [3]	Fully Loaded Chew [10]
Fully Loaded, Straight Pouches	F2	P	1.57 [3]	7.9 [3]	Fully Loaded Chew [10]
Fully Loaded, Peach Long Cut	F3	P	N/A	4.5 [3,4]	Fully Loaded Chew [10]
Fully Loaded, Berry Long Cut	F4	P	N/A	3.1 [3,4]	Fully Loaded Chew [10]
Black Buffalo, Straight Pouches	G1	P	1.11 [3]	5.1 [3]	Black Buffalo, Inc. [11]
Black Buffalo, Wintergreen Pouches	G2	P	1.01 [3]	4.3 [3]	Black Buffalo, Inc. [11]
Black Buffalo, Blood Orange Long Cut	G3	P	N/A	7.2 [3,4]	Black Buffalo, Inc. [11]
Black Buffalo, Peach Long Cut	G4	P	N/A	6.8 [3,4]	Black Buffalo, Inc. [11]
General Wintergreen White Portion	H1	T	0.96 [3]	6.9 [3]	Swedish Match [9]
Grizzly Wintergreen Long Cut	I1	T	N/A	10.0 [3,4]	American Snuff Co. [12]
CRP1.1	J1	T	1.01 [3]	8.1 [2]	CORESTA [13]
CRP2.1	J2	T	N/A	12.0 [2,4]	CORESTA [13]

[1] WGP = white granular powder; P = non-tobacco plant material; T = tobacco. [2] Information obtained from retail website or from product packaging. [3] Determined experimentally. [4] Nicotine content for long-cut products assumes a portion size of 1 g. [5] A wholly-owned subsidiary of Reynolds American, Inc. [27], based in Winston-Salem, NC, USA. [6] A wholly-owned subsidiary of the Altria Group, Inc. [28], based in Richmond, VA, USA. [7] A partnership between Swisher and Avema Pharma Solutions [29], based in Jacksonville, FL, USA. [8] A wholly-owned subsidiary of Turning Point Brands [30], based in Louisville, KY, USA. [9] Based in Stockholm, Sweden. [10] Based in Akron, OH, USA. [11] Based in Chicago, IL, USA. [12] A wholly-owned subsidiary of Reynolds American, Inc. [27], based in Memphis, TN, USA. [13] Based in Paris, France. CORESTA reference products were manufactured by and are distributed by the North Carolina State University Tobacco Analysis Service Laboratory (Raleigh, NC, USA) [31].

Table 6. IARC designations for the HPHCs of interest.

IARC Group	Compounds
1	NNK [10], NNN [10], Formaldehyde [9], Benzo[*a*]pyrene [32], Arsenic [11], Beryllium [11], Cadmium [11]
2A	Nitrite [33]
2B	Acetaldehyde [34], Crotonaldehyde [35], Cobalt [36], Lead [37], Nickel [38]
3	NAB [10], NAT [10], Chromium [38], Selenium [39]

Our own in-house methods (see Section 2.3) for the analysis of the selected HPHCs are fully validated and based upon the corresponding CORESTA recommended methods, with the exception of B[*a*]P, where a UHPLC method with fluorescence detection is used, and nitrite, which is based on Astoria-Pacific Method A181 (itself based on EPA method 353.2) [40]. The challenge presented by MONPs is that most HPHCs should be found in levels lower than what are considered typical in tobacco products. As a result, the limits of detection (LOD) and quantitation (LOQ) become more critical for each analytical method and may not be appropriate for the analysis of MONPs should the analytes be present. The 2021 CORESTA collaborative study [15] did not assess the suitability of the respective CRM LOQs, which were set with regard to the typical native levels found in tobacco products, and so this was an additional consideration in our study. The LOQs for our in-house methods are listed in Table 7, along with the LOQ values from the analogous CRM. Except for the metals and nitrite analysis methods, all of the LOQs for the validated methods are comparable to those listed in the analogous CRM. It is worth noting that CRMs are consensus methods that are evaluated by multiple laboratories before approval and publishing. As such, the method LOQs can be much higher than a particular laboratory may be capable of achieving in order to allow for differences in instrumentation and expertise across laboratories.

Table 7. Limits of detection and quantitation for Enthalpy's in-house analytical methods (see Section 2.3) and the analogous CRM quantitation limits (note: CRMs do not provide LODs).

Compound	Units	Enthalpy Analytical		CORESTA	
		LOD [1]	LOQ [1]	Method #	LOQ [1]
NAB	ng/g	0.51	3.3		3.8
NAT	ng/g	1.3	13	CRM 72 [24]	15
NNK	ng/g	1.3	13		15
NNN	ng/g	1.3	13		15
Nitrite	µg/g	0.033	0.10	CRM 36 [2] [41]	2 [3]
Benzo[*a*]pyrene	ng/g	0.060	0.18	CRM 82 [42]	0.15
Acetaldehyde	µg/g	0.063	0.090		0.10
Crotonaldehyde	µg/g	0.022	0.045	CRM 86 [25]	0.050
Formaldehyde	µg/g	0.050	0.090		0.10
Arsenic	ng/g	3.0	10		100–200 [4]
Beryllium	ng/g	0.25	2.5		100–200 [4]
Cadmium	ng/g	0.50	5.0		100–200 [4]
Chromium	ng/g	11	21	CRM 93 [26]	100–200 [4]
Cobalt	ng/g	0.60	5.0		100–200 [4]
Lead	ng/g	0.50	5.0		100–200 [4]
Nickel	ng/g	20	50		100–200 [4]
Selenium	ng/g	4.6	20		100–200 [4]

[1] Value is based on a nominal sample mass and will vary based on the actual sample mass used per replicate. [2] CRM 36 measures the nitrate content of a tobacco sample via reduction to nitrite and was shown in a 2016 collaborative study [43] to also be suitable for the analysis of nitrite. [3] LOQ value is based on the recommended minimum nitrate content of samples for analysis. [4] CRM 93 specifies a sample mass range of 0.5 to 1.0 g.

Using our in-house methods (see Section 2.3), we analyzed the selected products for the 17 HPHCs listed in Table 7 and, where necessary, nicotine. For ease of comparison, the results were initially calculated on a per-gram basis without correction for water content, a full summary of which can be found in Table S1 of the Supplementary Materials. The results were not corrected for any moisture content since our aim was to examine the consumer's potential exposure during use. Overall, the general trend observed followed what was expected, with the average levels measured in MONPs being lower than those measured in selected smokeless tobacco products. The trends for each analyte class are discussed in the following sections.

3.2. Nicotine

Since MONPs do not contain tobacco, nicotine is added during their manufacture. For the vast majority of the products, including most of those used in this study, the nicotine levels are reported on the packaging and range from 4 mg to 12 mg per portion (Table 5). This is comparable to the levels observed in the traditional smokeless products used as comparators (7 or 8 mg per portion for the snus products and 10 or 12 mg per portion for the long-cut products). Since some of the products selected for this study did not list their nicotine content on their packaging, they were assessed experimentally, and were found to exhibit comparable nicotine content to the other products being tested, ranging from 3.1 to 8.0 mg/portion.

3.3. Benzo[a]pyrene

Benzo[a]pyrene (B[a]P) is typically produced during combustion or any manufacturing process that may require heat (curing) [44]. Even if some MONPs are exposed to these types of processes, levels are still expected to be low. Of all the MONPs examined, only Fully Loaded, Berry Long Cut (F4) had detectable levels of B[a]P (1.27 ± 0.04 ng/g). This was far below those levels seen in the two long-cut STP comparators, which were 77.2 and 151 ng/g.

3.4. Nitrite and TSNAs

Nitrite is a precursor to TSNA formation [45,46] and is readily found in smokeless tobacco products [47]. Generally, the nitrite observed in MONPs was lower than those levels seen in STPs, except for one product (G2) that contained three-to-four times more nitrite than observed in the STPs. Given the low nitrite observed, it is not surprising that the levels of TSNAs observed for almost all of the MONPs in this study were below the LOQ or non-detectability threshold. The exceptions to this were the two Black Buffalo long-cut products (G3 and G4) that were found to contain NAT (18.5 ± 0.3 and 14.8 ± 1.3 ng/g, respectively), NNK (13.7 ± 0.4 ng/g in G3 only), and NNN (39.4 ± 1.3 and 32.4 ± 0.8 ng/g, respectively), shown in Figure 1. Both the NAT and NNK amounts were close to the respective LOQ values, while the NNN was close to three times the LOQ. In contrast to the MONPs, all four STPs were found to contain the four TSNAs at higher levels, particularly the CRP2.1 reference product (NAB: 9 to 274 ng/g; NAT: 125 to 4168 ng/g; NNK: 40 to 2104 ng/g; NNN: 177 to 3380 ng/g). While our analysis of MONPs shows that the LOQs associated with the method used are suitable for comparison to STPs, lower LOQs may be more beneficial when analyzing for regulatory reporting purposes. Adapting methods for trace TSNA analysis in ENDS e-liquid and aerosol may be a means for achieving this [48].

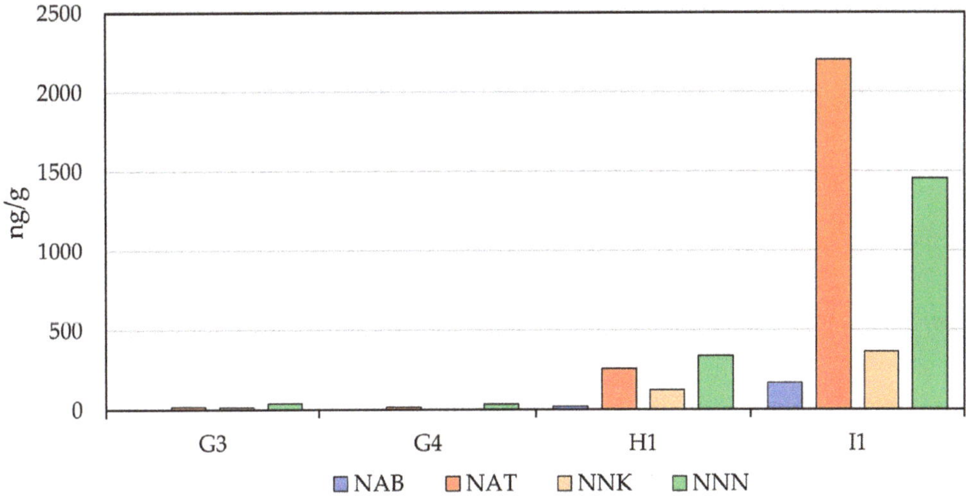

Figure 1. Comparison of the observed TSNA levels in the two long-cut Black Buffalo MONPs (G3 and G4) to the two commercial STPs (H1 and I1).

3.5. Carbonyls

One of the more interesting classes of compounds assessed in this study was carbonyls, for which there were distinct trends observed based solely on the filler composition of the MONP (Figure 2). The formaldehyde levels measured in all MONPs were comparable to those measured in the smokeless tobacco products tested (0.33 to 3.33 µg/g for MONPs and 0.78 to 3.64 µg/g for STPs). Two WGP-based MONPs, E1 and E2 (Zyn Nicotine Pouches), exhibited levels that were approximately three-to-four times those seen in STPs, being in the 10 to 14 µg/g range.

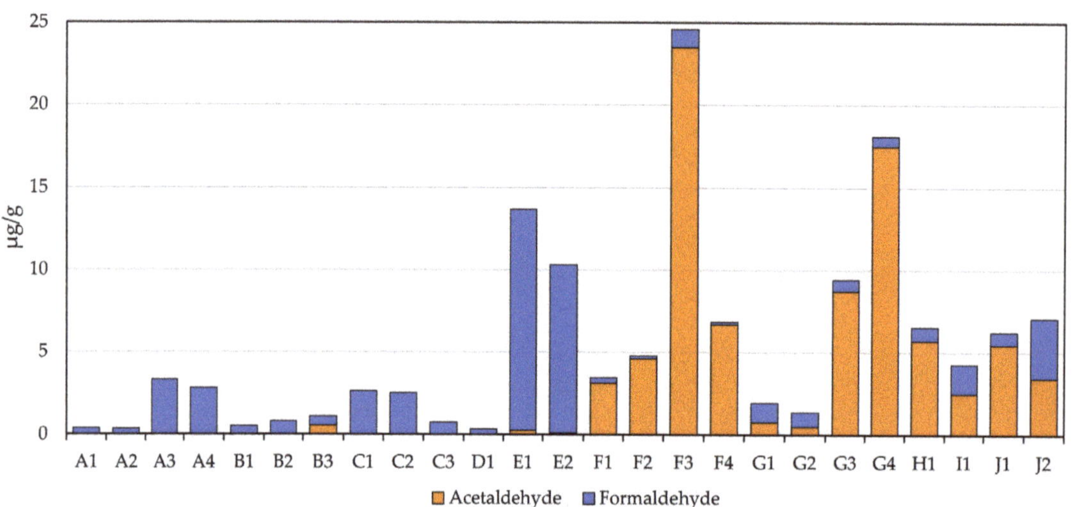

Figure 2. Observed levels of formaldehyde and acetaldehyde in the MONPs and four STPs.

Although the levels of formaldehyde were fairly consistent among all products screened, acetaldehyde appeared to be composition specific. The acetaldehyde levels measured in WGP-based products were typically below the LOQ or non-detectability except for one product, B3, that contained 0.56 ± 0.06 µg/g. Conversely, the acetaldehyde measured in the plant-based products ranged from lower to substantially higher than the levels observed in STPs, topping out at 23.5 µg/g versus 5.7 µg/g measured in the commercial snus product (H1).

Interestingly, the highest levels of acetaldehyde measured in MONP were observed in products F3 and G4, both of which happen to be peach flavored. It is also noteworthy that one of these products (G4) was the only product tested to have detectable crotonaldehyde, albeit just above the LOQ. Given that manufacturers will likely use the same base composition across their respective products, it is indicative of the elevated acetaldehyde levels observed being due to this particular flavorant.

3.6. Metals

Similar to carbonyls, there are clear differences in the metal analyte profiles of the two MONP types (Figure 3). WGP-based MONPs generally show much lower levels of metals, if present at all, than the plant- and tobacco-based products. Arsenic, beryllium, cadmium, and selenium were all either below the LOD or below the LOQ. Cobalt (4.5 to 10.4 ng/g) and lead (3.9 to 19.6) were all within six-times the LOQ (both 3.7 ng/g), except for lead in the Rogue Pouch products, C1 to C3 (77 to 83 ng/g), which were closer to the levels seen in the pouched STPs (approximately 107 ng/g). The more prevalent metals in WGP-MONPs were chromium (22.5 to 274 ng/g) and nickel (39.4 to 138 ng/g), though the levels were, for the most part, much lower than in the pouched STPS (Cr: 242 to 336 ng/g; Ni: 644 to 676 ng/g). The highest levels were seen for the Rogue Pouch products (Cr: 215 to 274 ng/g;

Ni: 92.5 to 138 ng/g), with two of the on! Nicotine Pouch products (B1 and B2) and Fré Lush also showing elevated Cr levels relative to the majority of the WGP-MONPs tested.

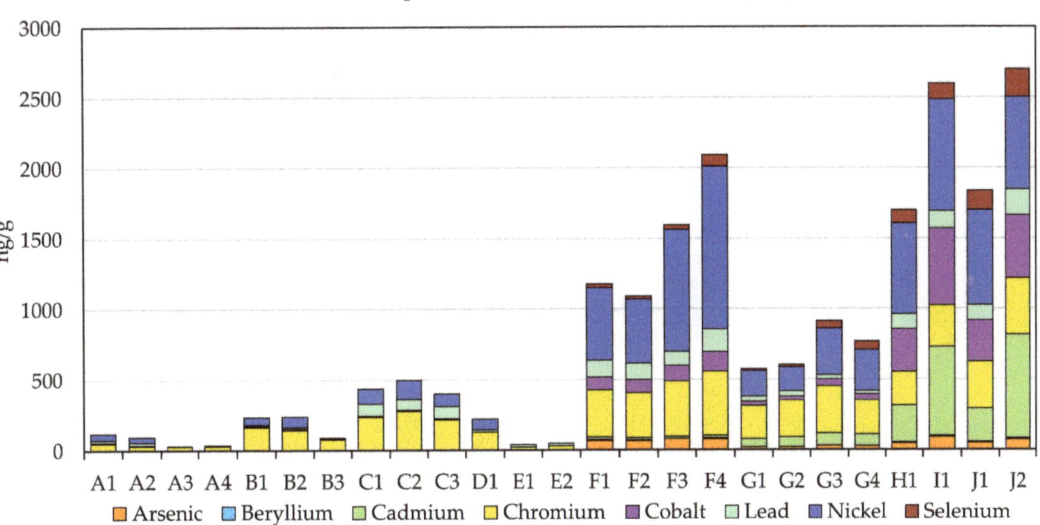

Figure 3. Observed levels of metals in the MONPs and four STPs.

In contrast to the WGP-based MONPs, the plant-based MONPs showed greater comparability with the tobacco-based products, though overall contained lower metal levels than the STPs with some noted exceptions. Beryllium was at or below the LOQ in all eight plant-based MONPs, and cadmium (16.0 to 86.0 ng/g) and cobalt (29.0 to 139 ng/g) were well below the levels in the STPs (Cd: 233 to 735 ng/g; Co: 293 to 546 ng/g). Arsenic (17.9 to 78.7 ng/g), chromium (233 to 456 ng/g), lead (21.7 to 159 ng/g), and selenium (12.4 to 81.8 ng/g) were observed at levels that were below or comparable to those in the STPs (As: 46.8 to 90.4 ng/g; Cr: 242 to 401 ng/g; Pb: 107 to 182 ng/g; Se: 93.5 to 197 ng/g). Nickel levels ranged from well below (175 ng/g in G2) to well above (1115 ng/g in F4) those in STPs (644 to 792 ng/g). Furthermore, there were clear differences between the two product brands tested, with the Fully Loaded brand generally containing higher metals levels than the Black Buffalo products for all analytes except cadmium and, for select products, selenium. This difference in levels may reflect the plants used to create the products and their respective uptake of metals from the soil and air. The generally higher levels in the Fully Loaded products may have been due to the use of kudzu root [49], compared with edible green leaves for Black Buffalo products [50], since a plant's root system is responsible for the absorption of nutrients from the surrounding soil. The kudzu plant was even shown to have utility for the lead phytoremediation of soils through rhizofiltration (root absorption) [51]. Finally, the metal content of the long-cut MONP formulations of the products were, with few exceptions, higher than their pouched counterparts. This may have been due to differences in the ingredients—the Fully Loaded long-cut formulation, for instance, contains molasses whereas the pouch formulation does not [49]—or it could simply have been due to batch variations in the base plant matrix used during manufacturing.

3.7. Exposure Assessment

One of the goals of reduced risk products is to be able to provide comparable nicotine delivery with reduced exposure to potential HPHCs. However, assessing potential exposure is not a trivial task. The frequency of use is directly associated with the potential exposure to HPHCs and can vary widely across the globe. For example, a study conducted in the United States examined adult usage of specific brands of snus and found the study subjects consumed an average of 3.3 pouches per day [52]. This is a relatively small amount when compared to the consumption habits found in Sweden where a survey conducted found the average daily consumption to be 11–12 g and 29–32 g for pouched snus and loose snus, respectively [53]. Due to this wide range in consumption habits, where available, the approximate portion size (pouch weight) was obtained and is provided in Table 5, providing context to potential usage. The observed HPHC analyte levels on a per-portion basis can also be found in Table S2 of the Supplementary Materials, and were calculated using the actual number of pouches analyzed per replicate (for long-cut products, a 1 g portion size was used). For the WGP-based MONPs, the HPHC amount per portion will be reduced relative to both the STP amounts and their own per-gram amount since they have portion weights below 0.7 g. The plant-based MONP pouches, on the other hand, are either comparable (Black Buffalo, G1 and G2) or higher (Fully Loaded, F1 and F2) than their per-gram amounts. Relative to the pouched STPs, there are few changes to the trends described above, the exceptions being for the heavier Fully Loaded pouch products in which acetaldehyde becomes comparable, and lead and nickel become higher per portion.

Overall, the MONPs appear to pose a much-reduced exposure risk compared to STPs, though the caveat here is that the sample size for STPs is very limited in this study and does not represent the full range of products available to the consumer. Ultimately, however, it will be the end user's consumption habits that determine their particular potential HPHC exposure risk.

4. Conclusions

The purpose of this work was to expand upon previous studies examining the potential HPHC content of modern oral nicotine products. With the push towards reducing the consumer's potential HPHC exposure, interest in the science surrounding modern oral nicotine products has increased. Previous work typically focused on assessing a single product brand, where this study sought to assess 25 unique products, which also includes two CORESTA reference products and two traditional smokeless tobacco products. Generally, products that are composed of powder-based materials displayed much lower levels of the HPHCs being assessed than those observed in the plant-based oral nicotine products. This trend was most evident in the metals and acetaldehyde analysis, and was likely due to a combination of higher native levels in the plants being used and any curing or manufacturing processes employed during production.

Although the levels reported here for MONPs are typically lower than the commercial and reference STPs analyzed, it should be noted that the STPs used in this study do not represent all smokeless tobacco products and these results should be viewed in that context. Further work is suggested to provide a more complete picture of the toxicant exposure risk to consumers with relation to all available marketed products. Potential areas for inclusion would be an examination into how the nicotine level may affect the HPHC amounts and a probe of potential batch-to-batch variability.

Supplementary Materials: The following supporting information can be downloaded at: https://www.mdpi.com/article/10.3390/separations9030065/s1, Table S1: Tabulated results on a per-gram basis for select HPHC analytes in the chosen MONPs and STP comparators; Table S2: Tabulated results on a per-portion basis for select HPHC analytes in the chosen MONPs and STP comparators; Raw data with calculations.xlsx.

Author Contributions: Conceptualization, J.J.J. and A.G.C.; methodology, J.J.J. and A.G.C.; formal analysis, J.J.J., A.G.C., and A.M.M.; investigation, J.J.J. and A.G.C.; resources, J.J.J., A.G.C., and A.M.M.; data curation, J.J.J. and A.G.C.; writing—original draft preparation, J.J.J. and A.G.C.; writing—review and editing, J.J.J., A.G.C., and A.M.M.; visualization, J.J.J., A.G.C. and A.M.M.; supervision, A.M.M.; project administration, J.J.J. All authors have read and agreed to the published version of the manuscript.

Funding: This study was supported by Enthalpy Analytical, LLC with internal funds. No external funding was received in support of this work. All commercial modern oral nicotine products and smokeless tobacco products were purchased from the retail market by Enthalpy Analytical staff.

Data Availability Statement: All data used in this work are available in the Supplementary Materials.

Conflicts of Interest: The authors declare the following financial/personal relationships that may be considered competing interests: The authors are employed by Enthalpy Analytical, LLC. Enthalpy Analytical, LLC is a commercial testing laboratory with a focus on nicotine-containing products and provides services to a wide range of clients including tobacco manufacturers and regulatory authorities. This study was funded internally by Enthalpy Analytical and no industry funding was accepted for the preparation of this manuscript. All products used in this study were purchased from the retail market by Enthalpy Analytical, LLC.

References

1. Family Smoking Prevention and Tobacco Control Act Table of Contents. Available online: https://www.fda.gov/tobacco-products/rules-regulations-and-guidance/family-smoking-prevention-and-tobacco-control-act-table-contents (accessed on 24 January 2022).
2. Food and Drug Administration Family Smoking Prevention and Tobacco Control Act—An Overview. Available online: https://www.fda.gov/tobacco-products/rules-regulations-and-guidance/family-smoking-prevention-and-tobacco-control-act-overview#smokeless (accessed on 4 April 2021).
3. Food and Drug Administration; US Department of Health and Human Services. Draft Guidance for Industry: Modified Risk Tobacco Product Applications; Availability; Agency Information Collection Activities; Proposed Collection; Comment Request. *Fed. Regist.* **2012**, *77*, 20026–20030.
4. Food and Drug Administration. Harmful and Potentially Harmful Constituents in Tobacco Products and Tobacco Smoke—Established List. *Fed. Regist.* **2012**, *77*, 20034–20037.
5. Wang, T.; Kenemer, B.; Tynan, M.; Singh, T.; King, B. Consumption of Combustible and Smokeless Tobacco—United States, 2000–2015. *MMWR Morb. Mortal. Wkly. Rep.* **2016**, *65*, 1357–1363. [CrossRef] [PubMed]
6. Timberlake, D.S. Are smokers receptive to using smokeless tobacco as a substitute? *Prev. Med.* **2009**, *49*, 229–232. [CrossRef] [PubMed]
7. Sami, M.; Timberlake, D.S.; Nelson, R.; Goettsch, B.; Ataian, N.; Libao, P.; Vassile, E. Smokers' perceptions of smokeless tobacco and harm reduction. *J. Public Health Policy* **2012**, *33*, 188–201. [CrossRef] [PubMed]
8. International Agency for Research on Cancer List of Classifications—IARC Monographs on the Identification of Carcinogenic Hazards to Humans. Available online: https://monographs.iarc.who.int/list-of-classifications/ (accessed on 12 April 2021).
9. International Agency for Research on Cancer. *Formaldehyde, 2-Butoxyethanol and 1-tert-Butoxypropan-2-ol*; International Agency for Research on Cancer: Lyon, France, 2006; ISBN 978-92-832-1288-1.
10. International Agency for Research on Cancer. *Smokeless Tobacco and Some Tobacco-Specific N-Nitrosamines*; World Health Organization: Geneva, Switzerland, 2006; ISBN 978-92-832-1289-8.
11. International Agency for Research on Cancer. *Arsenic, Metals, Fibres, and Dusts*; International Agency for Research on Cancer: Lyon, France, 2012; Volume 1989, ISBN 978-92-832-1320-8.
12. Borgerding, M.F.; Bodnar, J.A.; Curtin, G.M.; Swauger, J.E. The chemical composition of smokeless tobacco: A survey of products sold in the United States in 2006 and 2007. *Regul. Toxicol. Pharmacol.* **2012**, *64*, 367–387. [CrossRef] [PubMed]
13. Food and Drug Administration Modified Risk Orders. Available online: https://www.fda.gov/tobacco-products/advertising-and-promotion/modified-risk-orders (accessed on 11 April 2021).
14. Food and Drug Administration. FDA Grants First-Ever Modified Risk Orders to Eight Smokeless Tobacco Products. Available online: https://www.fda.gov/news-events/press-announcements/fda-grants-first-ever-modified-risk-orders-eight-smokeless-tobacco-products (accessed on 11 April 2021).
15. CORESTA TTPA Sub-Group. Technical Report 2021 Nicotine Pouches Collaborative Study. Available online: https://www.coresta.org/2021-nicotine-pouches-collaborative-study-35369.html (accessed on 24 January 2022).
16. Aldeek, F.; McCutcheon, N.; Smith, C.; Miller, J.H.; Danielson, T.L. Dissolution Testing of Nicotine Release from OTDN Pouches: Product Characterization and Product-to-Product Comparison. *Separations* **2021**, *8*, 7. [CrossRef]
17. Wagner, K.A.; Brown, A.P.; Jin, X.C.; Sharifi, M.; Lopez, V.F.; Ballentine, R.M.; Melvin, M.S.; Mcfarlane, C.B.; Morton, M.J.; Danielson, T.L. Characterization of on! ® nicotine pouches—Part 1: HPHCs. In Proceedings of the SRNT 26th Annual Meeting, New Orleans, LA, USA, 11–14 March 2020.
18. Bishop, E.; East, N.; Bozhilova, S.; Santopietro, S.; Smart, D.; Taylor, M.; Meredith, S.; Baxter, A.; Breheny, D.; Thorne, D.; et al. An approach for the extract generation and toxicological assessment of tobacco-free 'modern' oral nicotine pouches. *Food Chem. Toxicol.* **2020**, *145*, 111713. [CrossRef]
19. East, N.; Bishop, E.; Breheny, D.; Gaca, M.; Thorne, D. A screening approach for the evaluation of tobacco-free 'modern oral' nicotine products using Real Time Cell Analysis. *Toxicol. Rep.* **2021**, *8*, 481–488. [CrossRef] [PubMed]
20. Azzopardi, D.; Liu, C.; Murphy, J. Chemical characterization of tobacco-free "modern" oral nicotine pouches and their position on the toxicant and risk continuums. *Drug Chem. Toxicol.* **2021**, 1–9. [CrossRef] [PubMed]
21. *ASTM D1193-06(2018)*; Standard Specification for Reagent Water. ASTM International. Available online: https://www.astm.org/d1193-06r18.html (accessed on 24 January 2022).

22. Centers for Disease Control and Prevention (CDC); US Department of Health and Human Services. Notice Regarding Revisions to the Laboratory Protocol To Measure the Quantity of Nicotine Contained in Smokeless Tobacco Products Manufactured, Imported, or Packaged in the United States. *Fed. Regist.* **2009**, *74*, 609–768.

23. Cooperation Centre for Scientific Research Relative to Tobacco CORESTA Recommended Method No. 62 Determination of Nicotine in Tobacco and Tobacco Products by Gas Chromatographic Analysis. Available online: https://www.coresta.org/determination-nicotine-tobacco-and-tobacco-products-gas-chromatographic-analysis-29185.html (accessed on 21 February 2022).

24. Cooperation Centre for Scientific Research Relative to Tobacco CORESTA Recommended Method No. 72 Determination of Tobacco Specific Nitrosamines in Tobacco and Tobacco Products by LC-MS/MS. Available online: https://www.coresta.org/determination-tobacco-specific-nitrosamines-tobacco-and-tobacco-products-lc-msms-29195.html (accessed on 24 January 2022).

25. Cooperation Centre for Scientific Research Relative to Tobacco CORESTA Recommended Method No. 86 Determination of Select Carbonyls in Tobacco and Tobacco Products by UHPLC-MS/MS. Available online: https://www.coresta.org/determination-select-carbonyls-tobacco-and-tobacco-products-uhplc-msms-30991.html (accessed on 24 January 2022).

26. Cooperation Centre for Scientific Research Relative to Tobacco CORESTA Recommended Method No. 93 Determination of Selected Metals in Tobacco Products by ICP-MS. Available online: https://www.coresta.org/determination-selected-metals-tobacco-products-icp-ms-33784.html (accessed on 24 January 2022).

27. Brands. Available online: https://www.reynoldsamerican.com/brands (accessed on 25 January 2022).

28. Our Companies—Altria. Available online: https://www.altria.com/en/about-altria/our-companies (accessed on 25 January 2022).

29. Who is Rogue Holdings, LLC?—Rogue NOD. Available online: https://help.roguenicotine.com/hc/en-us/articles/360054562234-Who-is-Rogue-Holdings-LLC- (accessed on 25 January 2022).

30. Turning Point Brands—Our Company—Company Timeline. Available online: https://www.turningpointbrands.com/our-company/company-timeline/default.aspx (accessed on 25 January 2022).

31. Smokeless Tobacco Reference Products. Available online: https://strp.wordpress.ncsu.edu/ (accessed on 24 February 2022).

32. International Agency for Research on Cancer. *Some Non-Heterocyclic Polycyclic Aromatic Hydrocarbons and Some Related Exposures*; International Agency for Research on Cancer: Lyon, France, 2005; ISBN 978-92-832-1292-8.

33. International Agency for Research on Cancer. *Ingested Nitrate and Nitrite, and Cyanobacterial Peptide Toxins*; International Agency for Research on Cancer: Lyon, France, 2010; ISBN 978-92-832-1294-2.

34. International Agency for Research on Cancer. *Re-Evaluation of Some Organic Chemicals, Hydrazine and Hydrogen Peroxide (Part 1, Part 2, Part 3)*; World Health Organization: Geneva, Switzerland; International Agency for Research on Cancer: Lyon, France, 1999; ISBN 978-92-832-1271-3.

35. International Agency for Research on Cancer. *Acrolein, Crotonaldehyde, and Arecoline*; International Agency for Research on Cancer: Lyon, France, 2021; ISBN 978-92-832-0195-3.

36. International Agency for Research on Cancer. *Chlorinated Drinking-Water; Chlorination By-Products; Some Other Halogenated Compounds; Cobalt and Cobalt Compounds*; IARC: Lyon, France, 1991; ISBN 978-92-832-1252-2.

37. International Agency for Research on Cancer. *Some Metals and Metallic Compounds*; International Agency for Research on Cancer: Lyon, France, 1980; Volume 23, ISBN 978-92-832-1223-2.

38. International Agency for Research on Cancer. *Chromium, Nickel and Welding*; IARC: Lyon, France, 1990; ISBN 978-92-832-1249-2.

39. International Agency for Research on Cancer. *Some Aziridines, N-, S- and O-Mustards and Selenium*; World Health Organization: Geneva, Switzerland, 1975; ISBN 978-92-832-1209-6.

40. O'Dell, J.W. *U.S. Environmental Protection Agency Method 353.2, Revision 2.0: Determination of Nitrate-Nitrite Nitrogen by Automated Colorimetry*; US Environmental Protection Agency: Cincinnati, OH, USA, 1993.

41. Cooperation Centre for Scientific Research Relative to Tobacco CORESTA Recommended Method No. 36 Determination of Nitrate in Tobacco and Smokeless Tobacco Products by Reduction to Nitrite and Continuous Flow Analysis. Available online: https://www.coresta.org/determination-nitrate-tobacco-and-smokeless-tobacco-products-reduction-nitrite-and-continuous-flow (accessed on 24 January 2022).

42. Cooperation Centre for Scientific Research Relative to Tobacco CORESTA Recommended Method No. 82 Determination of Benzo[a]pyrene in Tobacco Products by GC-MS. Available online: https://www.coresta.org/determination-benzoapyrene-tobacco-products-gc-ms-29897.html (accessed on 24 January 2022).

43. CORESTA TTPA Sub-Group Technical Repor Determination of Nitrite and Nitrate in Smokeless Tobacco Products by Ion Chromatography and Continuous Flow Analysis—2016 Collaborative Study. Available online: https://www.coresta.org/determination-nitrite-and-nitrate-smokeless-tobacco-products-ion-chromatography-and-continuous-flow (accessed on 24 January 2022).

44. McAdam, K.G.; Faizi, A.; Kimpton, H.; Porter, A.; Rodu, B. Polycyclic aromatic hydrocarbons in US and Swedish smokeless tobacco products. *Chem. Cent. J.* **2013**, *7*, 151. [CrossRef] [PubMed]

45. Nestor, T.B.; Gentry, J.S.; Peele, D.M.; Riddick, M.G.; Conner, B.T.; Edwards, M.E. Role of Oxides of Nitrogen in Tobacco-Specific Nitrosamine Formation in Flue-Cured Tobacco. *Beitr. Table Int.* **2014**, *20*, 467–475. [CrossRef]

46. Burton, H.R.; Dye, N.K.; Bush, L.P. Relationship between Tobacco-Specific Nitrosamines and Nitrite from Different Air-Cured Tobacco Varieties. *J. Agric. Food Chem.* **1994**, *42*, 2007–2011. [CrossRef]

47. Andersen, R.A.; Burton, H.R.; Fleming, P.D.; Hamilton-Kemp, T.R. Effect of Storage Conditions on Nitrosated, Acylated, and Oxidized Pyridine Alkaloid Derivatives in Smokeless Tobacco Products. *Cancer Res.* **1989**, *49*, 5895–5900.

48. Moldoveanu, S.C.; Zhu, J.; Qian, N. Analysis of Traces of Tobacco-Specific Nitrosamines (TSNAs) in USP Grade Nicotine, E-Liquids, and Particulate Phase Generated by the Electronic Smoking Devices. *Beiträge zur Table Int. to Tob. Res.* **2017**, *27*, 86–96. [CrossRef]
49. FAQs—Frequenty Asked Questions—Fully Loaded LLC. Available online: https://fullyloadedchew.com/pages/faq (accessed on 24 January 2022).
50. FAQs On Dipping Black Buffalo Long Cut & Pouches. Available online: https://blackbuffalo.com/pages/faq (accessed on 24 January 2022).
51. Schwarzauer-Rockett, K.; Al-Hamdani, S.H.; Rayburn, J.R.; Mwebi, N.O. Utilization of kudzu as a lead phytoremediator and the impact of lead on selected physiological responses. *Can. J. Plant Sci.* **2013**, *93*, 951–959. [CrossRef]
52. Caraway, J.W.; Chen, P.X. Assessment of Mouth-Level Exposure to Tobacco Constituents in U.S. Snus Consumers. *Nicot. Tob. Res.* **2013**, *15*, 670–677. [CrossRef] [PubMed]
53. Digard, H.; Errington, G.; Richter, A.; McAdam, K. Patterns and behaviors of snus consumption in Sweden. *Nicot. Tob. Res. Off. J. Soc. Res. Nicot. Tob.* **2009**, *11*, 1175–1181. [CrossRef] [PubMed]

separations

Article

Determination of Formaldehyde Yields in E-Cigarette Aerosols: An Evaluation of the Efficiency of the DNPH Derivatization Method

Xiaohong C. Jin, Regina M. Ballentine, William P. Gardner, Matt S. Melvin, Yezdi B. Pithawalla, Karl A. Wagner, Karen C. Avery and Mehran Sharifi *

Center For Research & Technology, Altria Client Services LLC, 601 East Jackson Street, Richmond, VA 23219, USA; Xiaohong.Jin@altria.com (X.C.J.); Regina.M.Ballentine@altria.com (R.M.B.); William.P.Gardner@altria.com (W.P.G.); Matt.S.Melvin@altria.com (M.S.M.); Yezdi.B.Pithawalla@altria.com (Y.B.P.); Karl.A.Wagner@altria.com (K.A.W.); Karen.C.Avery@altria.com (K.C.A.)
* Correspondence: Mehran.Sharifi@altria.com; Tel.: +1-(804)-335-2062

Abstract: Recent reports have suggested that (1) formaldehyde levels (measured as a hydrazone derivative using the DNPH derivatization method) in Electronic Nicotine Delivery Systems (ENDS) products were underreported because formaldehyde may react with propylene glycol (PG) and glycerin (Gly) in the aerosol to form hemiacetals; (2) the equilibrium would shift from the hemiacetals to the acetals in the acidic DNPH trapping solution. In both cases, neither the hemiacetal nor the acetal would react with DNPH to form the target formaldehyde hydrazone, due to the lack of the carbonyl functional group, thus underreporting formaldehyde. These reports were studied in our laboratory. Our results showed that the aerosol generated from formaldehyde-fortified e-liquids provided a near-quantitative recovery of formaldehyde in the aerosol, suggesting that if any hemiacetal was formed in the aerosol, it would readily hydrolyze to free formaldehyde and, consequently, form formaldehyde hydrazone in the acidic DNPH trapping solution. We demonstrated that custom-synthesized Gly and PG hemiacetal adducts added to the DNPH trapping solution would readily hydrolyze to form the formaldehyde hydrazone. We demonstrated that acetals of PG and Gly present in e-liquid are almost completely transferred to the aerosol during aerosolization. The study results demonstrate that the DNPH derivatization method allows for an accurate measurement of formaldehyde in vapor products.

Keywords: e-cigarette; e-liquid; aerosol; 2,4-DNPH derivatization; formaldehyde; "hidden formaldehyde"; formaldehyde-containing hemiacetal/acetal adducts

Citation: Jin, X.C.; Ballentine, R.M.; Gardner, W.P.; Melvin, M.S.; Pithawalla, Y.B.; Wagner, K.A.; Avery, K.C.; Sharifi, M. Determination of Formaldehyde Yields in E-Cigarette Aerosols: An Evaluation of the Efficiency of the DNPH Derivatization Method. *Separations* **2021**, *8*, 151. https://doi.org/10.3390/separations8090151

Academic Editors: Fadi Aldeek and Beatriz Albero

Received: 13 August 2021
Accepted: 7 September 2021
Published: 13 September 2021

Publisher's Note: MDPI stays neutral with regard to jurisdictional claims in published maps and institutional affiliations.

1. Introduction

Formaldehyde (FA) is classified as a Group 1 carcinogen in humans by the International Agency for Research on Cancer (IARC) [1]. Formaldehyde is a common indoor air pollutant due to its ubiquitous use in the production of various industrial products [2]. Thus, one source of human exposure to formaldehyde is its release from household products made using formaldehyde or containing formaldehyde-releaser compounds that are placed in poorly ventilated areas [3,4]. Cigarette smoke is reported as another common source of exposure to formaldehyde, which is formed as a byproduct of the combustion process of tobacco [3]. Regulations for reporting formaldehyde yields in cigarette smoke are enacted by different regulatory authorities [5,6]. More recently, the Food and Drug Administration (FDA) cataloged a list of "Harmful and Potentially Harmful Constituents" (HPHCs) of tobacco products, which includes formaldehyde [7,8]. The FDA's Guidance to Industry regarding the submission of Premarket Tobacco Applications for Electronic Nicotine Delivery Systems (ENDS) also includes formaldehyde on the list of constituents "that would potentially cause health hazards depending on the level, absorption, or interaction with other constituents" [9].

Formaldehyde yields reported in machine-generated smoke from commercially available cigarettes vary (~10–70 µg/cigarette depending on the tobacco blend, cigarette design, and intensity of the smoking conditions [10–12]). Formaldehyde has also been reported in e-cigarette emissions [13–16]. The formation of formaldehyde in e-cigarette vapor is mainly attributed to the thermal degradation of propylene glycol (PG) and glycerol (Gly) and select flavoring agents [14–21]. Though typically at much lower levels than in tobacco smoke [22,23], a wide discrepancy in formaldehyde levels (0.5–50 µg/puff) has been reported in emissions from across commercially available e-cigarette products. The formaldehyde formation in e-cigarette aerosol is indeed related to the aerosolization efficiency of e-cigarette devices, which depends mainly on vaporizer physical and electronic design (temperature control, air flow, pressure drop, etc.), as well as the quality of materials used in manufacturing the device (heating coil element, liquid-containing cartridge, and wick) [14]. Other factors that influence the formation of formaldehyde include e-liquid components (propylene glycol, glycerol, and some flavorings), the propensity of the device to "dry-puff," thereby resulting in higher vaporization temperatures, and operating parameters of the device (voltage and puffing strength) [13–16,18,20,24–27].

For instance, a drastic increase in formaldehyde emission rate (from 0.1 to 30 µg/puff) was observed by increasing the voltage applied to a single-coil device from 3.3 to 5 V [28]. Gillman et al. reported [14] that the power intensity applied on the coil is not the sole factor affecting formaldehyde emission rates and that general device design characteristics such as coil position (top or bottom), single or dual coil-head, and coil resistance play a significant role in the formaldehyde generation process that occurs during aerosolization. The authors [14] further reported that an increase in power from 5 to 9 W in a single bottom-coil induced a drastic 70-fold increase in formaldehyde emission rate as opposed to a 6-fold increase observed using a single top-coil tank.

Due to its high reactivity, its low molecular mass, and the lack of a strong chromophore, a direct determination of formaldehyde in smoke or e-cigarette aerosol is typically achieved via a derivatization step. The conventional derivatization methodology is based on an acid-catalyzed condensation reaction between carbonyl compounds and 2,4-dinitrophenylhydrazine (2,4-DNPH). This method is described in several standardized methods, including US-EPA, NIOSH, and ISO, and has been widely used outside of nicotine products. The reaction proceeds by nucleophilic addition of the hydrazine functionality to the carbonyl compound, followed by elimination of water to form the corresponding hydrazone (Scheme 1).

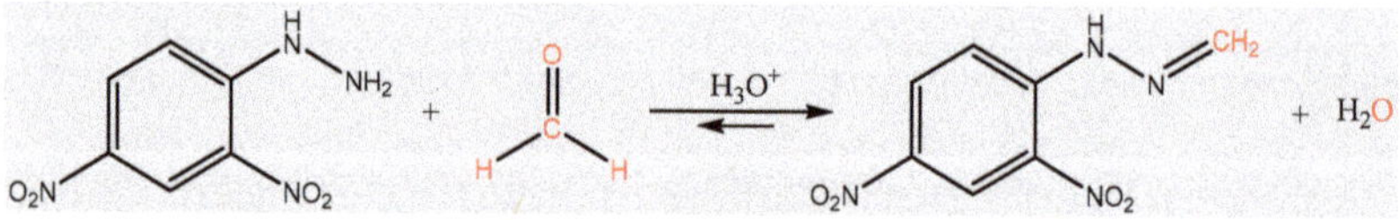

Scheme 1. Derivatization of formaldehyde by 2,4-DNPH. The red color is used to visualize the condensation site of the methyl moiety of formaldehyde within the FA-hydrazone molecule.

The DNPH derivatization approach for the determination of formaldehyde in cigarette smoke has been developed and validated by multiple organizations, including CORESTA (Centre de Coopération pour les Recherches Scientifiques Relatives au Tabac) [29], Health Canada [30], and International Organization for Standardization (ISO) [31]. The conventional DNPH method has been widely utilized over the past decades in the tobacco industry and at independent analytical testing facilities for measuring formaldehyde yields in both conventional and electronic cigarettes.

The application of the conventional DNPH derivatization methodology for trapping and quantifying formaldehyde in e-liquids and e-cigarette aerosols presented challenges, mainly due to formaldehyde's extremely low concentration [22,23], its endogenous levels in laboratory air, and its background level in DNPH reagent [26]. In order to overcome these obstacles, modifications to the existing method for analyzing cigarette smoke with respect to sample collection (i.e., use of DNPH-coated adsorption cartridges in lieu of impingers) and an alternative derivatization method (i.e., PFBHA) were undertaken by different laboratories using various analytical techniques (i.e., HPLC–DAD, LC–MS/MS, SPME/GC–MS, and GC–MS) [14,22,24,32–34].

Despite the widespread use of DNPH derivatization for the analysis of carbonyls in e-cigarette aerosol, in a paper published in 2017 [35], the authors theorized that the DNPH method significantly underestimates formaldehyde levels produced in e-cigarette aerosol. This theory was based on the assumption that formaldehyde-hemiacetal adducts, labeled "hidden formaldehyde," are formed in aerosol by the reversible addition of glycerol (primary hydroxyl group) and/or propylene glycol to the formaldehyde carbonyl functional group during aerosolization. The formaldehyde-hemiacetal (FA-hemiacetal) adduct(s) could then undergo an irreversible dehydration reaction catalyzed by the acidity of the DNPH trapping solution or silica sorbent (DNPH cartridge) to form two cyclic acetal isomers (Figure 1) [36]. The authors stated that the sequestrated formaldehyde portion in the form of hemiacetal (FA-hemiacetal) and/or acetal (FA-acetal) would not react with DNPH to form formaldehyde hydrozone and, thus, would not be measurable by the UV or MS detection used in the method, and therefore, the DNPH derivatization is not fit to measure total formaldehyde yields in e-cigarette aerosol, due to the inaccurate estimation of a user's exposure to formaldehyde [35]. They labeled this phenomenon as "hidden formaldehyde".

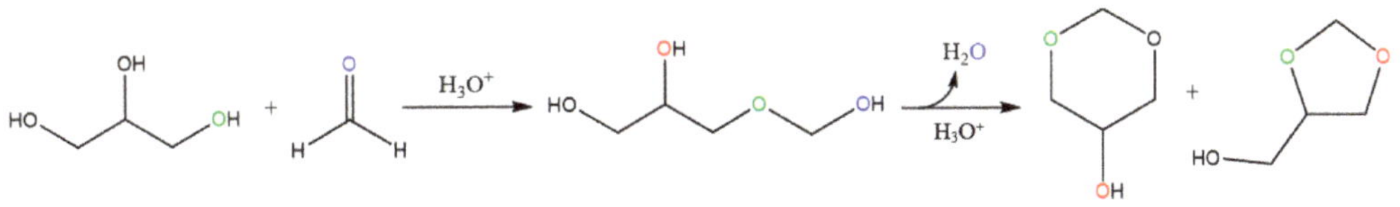

Figure 1. Formation of FA-glycerol hemiacetal (Gly-HA) and cyclic acetals (Gly-A). The colors are used to visualize the inclusion sites of various oxygen atoms in the reaction products.

Jensen and co-authors [36] estimated that an e-cigarette user vaping at a rate of 3 mL per day would inhale 14.4 ± 3.3 mg of formaldehyde per day in formaldehyde-hemiacetals and extrapolated their results to suggest an estimated increase in lifetime cancer risk by up to 15 fold higher to the risk for regular smokers. However, this study was criticized for being conducted under "unrealistic" user conditions and therefore misleading with respect to real user exposure to formaldehyde [28,37,38]. In response to the Jensen et al. study report [36], several letters were addressed to the journal editor requesting the retraction of the paper based on "fundamental flaws in the experimental and cancer risk calculations" [37]. Additional studies were conducted to replicate Jensen et al.'s findings using the same (or similar) atomizer, e-liquid, and operating conditions, which concluded that under "realistic" use conditions, formaldehyde yields in e-cigarette emissions are much lower than levels measured in cigarette smoke [28,39].

This paper describes the results from an evidence-based analytic approach to provide an objective assessment of the DNPH method performance with respect to formaldehyde quantification in e-cigarette emissions. A series of experiments were conducted to elucidate the reactivity of formaldehyde-containing acetal and hemiacetal adducts (listed in Figure 2) in the presence of an acidic DNPH derivatization solution. Additional experiments were conducted to determine whether acetals were formed during the aerosolization process or by intramolecular conversion of the hemiacetals to the cyclic acetals in acidic DNPH trapping solution. The analytical procedures used for analysis of formaldehyde, formaldehyde-containing hemiacetals (Glyα-HA and PGα-HA), and formaldehyde-containing acetals (Gly-A and PG-A) in e-liquid and/or DNPH trapping solution are described in the upcoming section.

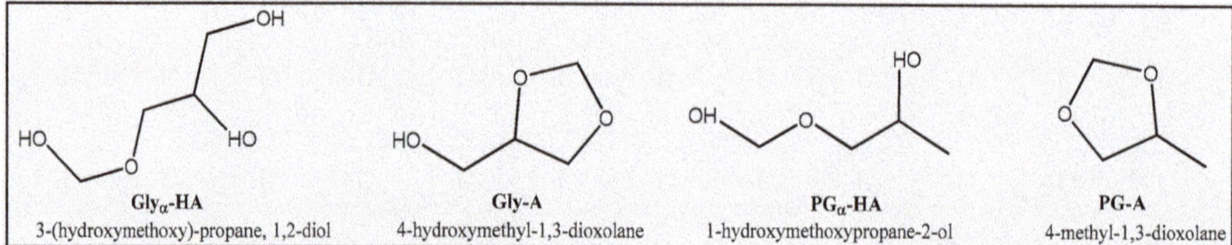

Figure 2. Formaldehyde-containing hemiacetal (Glyα-HA and PGα-HA) and acetal (Gly-A and PG-A) adducts.

2. Materials and Methods

Test Products. Two types of rechargeable e-cigarette devices (cig-a-like with disposable pre-filled cartridges and self-contained pod systems with refills) were purchased at retail locations in the 2018–2019 timeframe. All devices and flavors used for this study are listed in Table 1.

Table 1. Market test products.

Device Type Brand ID	Flavor ID	Nicotine by Weight (%)	Product Code
Cig-a-like_A	E1	1.5	CAE1
Cig-a-like_B	E2	4.8	CBE2
	E3		CBE3
Cig-a-like_C	E4	2.4	CCE4
	E5		CCE5
	E6		CCE6
Cig-a-like_D	E7	2.4	CDE7
	E8	3.5	CDE8
Pod_E	E9	2.4	PEE9
Pod_F	E10	5.0	PFE10
Pod_G	E11	3.0	PGE11

A reference formulation (15% water, 2.5% nicotine by weight (NBW) in a 50/50 mixture of PG and Gly) was also prepared in our laboratory in order to investigate the possible formation and transfer of formaldehyde hemiacetal and acetal adducts. Aerosols were generated using empty Cig-a-like commercial E cartridges (provided by a manufacturer) that were filled with either commercial or fortified e-liquid.

Chemicals and Reagents. Certified formaldehyde-DNPH hydrazone (FA-DNPH) solution in acetonitrile (700.2 μg/mL corresponding to 100 μg/mL in formaldehyde) was supplied by AccuStandard (New Haven, CT, USA). Deuterium-labeled formaldehyde-d3-3,5,6-DNPH (FA-d3-DNPH) was purchased from CDN Isotopes (Pointe-Claire, QC, Canada) and labeled as ≥99.7% pure.

The following formaldehyde-containing hemiacetal adducts, 3-(hydroxymethoxy)-propane, 1,2-diol (Glyα-HA, neat material, ≥98% pure by NMR), and 1-hydroxymethoxypropane-2-ol (PGα-HA, 50–60% pure by NMR), were custom-synthesized by Chemische Laboratorien Dr. Sönke Petersen (Worms, Germany). Glycerol formal (Gly-A) and 4-methyl-1,3-dioxolane (PG-A) were supplied by TCI (Portland, OR, USA) and Millipore Sigma (Milwaukee, WI, USA) and labeled as ≥98% pure. Certified deuterium-labeled benzene (d6-benzene) and 2,3-hexandione, used as internal standards for analysis of acetal adducts by GC–MS, were purchased from Restek (Bellefonte, PA, USA) and Alfa Aesar (Tewksbury, MA, USA), respectively.

The 2,4-dinitrophenylhydrazine hydrochloride salt (DNPH, HCl) was purchased from TCI America (Portland, OR, USA) and was labeled ≥98% pure. An acidified solution of DNPH (19 mM) was prepared in-house by dissolving purchased DNPH in acetonitrile containing 1.5% of an aqueous perchloric acid (1.82 M) solution [29]. The derivatization reagent solution was filtered and analyzed by HPLC–MS to ensure that the FA background was ≤0.05 μg/mL. A 60% solution of perchloric acid (0.6 M) was supplied by EMD

Millipore (Billerica, MA, USA). Acetonitrile and dichloromethane were distilled-in-glass grade. Type I reagent water was generated in-house as per American Society for Testing and Materials D1193 standard specification.

Sample Generation. E-cigarette aerosol was generated on a Borgwaldt LX20 linear smoking machine (Borgwaldt, Hamburg, Germany). The aerosol yields were obtained by collecting 50 puffs using a square-wave puff profile with a 5 s puff duration, 30 s puff interval, and a 55 mL puff volume.

The aerosol collection system for formaldehyde puffing experiments included a 44 mm-glass fiber filter pad and a 215 mm × 30 mm O.D. Drechsel-type bottle container (Prism Research Glass, Raleigh, NC, USA) enclosing the derivatization reagent (30 mL of DNPH solution). The aerosol was drawn through the filter pad followed by the impinging trap. Any formaldehyde collected on the filter pad was extracted/derivatized by adding the filter pad to the DNPH trapping solution. One milliliter of aerosol extract was then transferred to an amber autosampler vial containing 25 μL of pyridine (to stop the derivatization), and then 50 μL of FA-d3-DNPH solution (2 μg/mL) was added. The sample was then analyzed using an in-house-validated UPLC–MS detection method [40].

FA-DNPH Determination. The FA-DNPH content in e-liquid was determined by extracting 100 mg of the sample in 30 mL of DNPH reagent, which was left at room temperature for 5 min after mixing the reactants to allow the reactions to be completed. The reaction was stopped by adding pyridine and the sample was subject to UPLC-MS (Waters, Milford, MA, USA) analysis, as described later.

Acetal Determination. For the acetal puffing experiments, the aerosol was collected on a 44 mm glass fiber filter pad mounted in series with an impinging glassware containing dichloromethane (20 mL) and cooled in an ice bath (0 °C) to minimize the loss of trapping solvent. After aerosol generation, the filter pad and the impinger content were combined, and then 2 mL of type 1 water and the internal standard were added (d6-benzene or 2,3-hexandione). The mixture was vortexed for 20 min and acetal adducts were extracted by liquid-phase extraction (LPE) into the organic phase (20 mL of dichloromethane). An aliquot of dichloromethane was then analyzed by GC-MS (Agilent, Santa Clara, CA, USA), as described later.

The FA-acetal levels in the e-cigarette liquids were determined by adding the internal standard (d6-benzene or 2,3-hexandione) directly to 250 mg of the sample, which was then extracted in a type 1 water:dichloromethane mixture (2:20, v/v). The mixture was vortexed for 20 min and an aliquot of the organic phase containing acetal adducts was then subject to GC-MS analysis.

Analytical Methods. The analysis of FA-DNPH was conducted by UPLC-MS using a Waters Acquity UPLC system equipped with a binary pump, autosampler, and a TQ-S-Micro triple quadrupole mass analyzer with an electrospray ionization interface (ESI) (Waters, Milford, MA, USA). The UPLC separation was performed on a reversed-phase analytical column (Acquity UPLC BEH® C18, 2.1 × 50 mm, particle size 1.7 μm) from Waters (Milford, MA, USA) using a mixture of 10 mM of ammonium acetate/methanol (98:2 v/v) (mobile phase A) and a mixture of acetonitrile/1-propanol (90/10 v/v) (mobile phase B). The gradient program was as follows: initially constant at 65% A and 35% B for 2 min, the composition was then changed to 40% A and 60% B by a linear gradient occurring within 2 min, and then restored to the initial composition within 2.7 min and kept constant for 5 min. The flow rate was constant at 0.5 mL/min and the column temperature set to 45 °C. The ESI mass spectra for FA-DNPH and FA-d3-DNPH were acquired in negative ionization mode by monitoring their respective [M-H] molecular species (m/z 209 and m/z 212, respectively). The capillary and cone voltages were set at 0.65 kV and −32 V, respectively. The source block desolvation temperature was set to 450 °C and the source temperature was set at 150 °C. The method was validated based upon the 2005 International Conference on Harmonization (ICH) guideline "Validation of Analytical Procedures: Text and Methodology Q2(R1)" [41]. Repeatability each day was 3–12.7% of RSD for the analysis of 5 independently prepared replicate samples. Over the course of 3 days, the method variability (intermediate precision) within samples ranged from 1.66% to 14.8% %RSD. Selected ion monitoring is a specific detection technique and no interference peaks in the samples were observed. Accuracies were 90.7–106%. The limit of quantitation (LOQ) is defined as the lowest quantifiable level of formaldehyde such that the signal-to-noise ratio (S/N) is 10. The concentration of formaldehyde in the calibration standards ranged from 0.01 to 3.8 µg/mL with R^2 greater than 0.995 and percent deviation values (residuals) for all calibration levels ≤15% from their respective theoretical values using a linear calibration model. The LOQ was 3 µg/g for liquid and 0.3 µg/collection (corresponding to 1 µg/g of consumed e-liquid). Furthermore, the aerosol collection trapping efficiency study indicated that over 99% of formaldehyde was collected with one pad and one impinger, while formaldehyde was not observed in the 2nd impinger.

The yield of acetal adducts in aerosol emissions was determined by GC-MS. The GC–MS system consisted of an Agilent 7980 gas chromatograph system coupled with a 5977A MS single quadrupole mass analyzer, equipped with a conventional electron ionization (EI) source. The chromatographic separation was conducted on a Rtx®-624 fused-silica capillary column (30 m × 0.25 mm × 1.4 µm film thickness) crossbonded with (6% cyanopropylphenyl/49% dimethylpolysiloxane phase), purchased from Restek (Bellefonte, PA, USA). An optimized GC oven temperature program was established where the oven temperature was initially held at 50 °C for 2 min, ramped to 75 °C at a rate of 10 °C/min, and then ramped to 235 °C at 10 °C/min and held for 3 min. Helium was used as the carrier gas at a constant flow rate of 1.4 mL/min. The GC injector was set to 230 °C, and 2 µL aliquots of samples were injected in splitless mode. The EI mass spectra for Gly-A, PG-A, d6-benzene, and 2,3-hexandione were acquired in EI mode (−70 eV) by monitoring their respective $[M^+]$ molecular species (m/z 104, m/z 88, m/z 84, and m/z 114, respectively) with the dwell time value set at 50 milliseconds. The ion source and quadrupole temperatures were set at 230 °C and 150 °C, respectively. The concentrations of PG-A and Gly-A adducts in calibration standard solutions ranged between 0.01 and 2 µg/mL. The LOQ was determined as 0.8 µg/g of e-liquid and 2 µg/collection (corresponding to 0.8 µg/g of consumed e-liquid).

Analytical experiments. The investigatory approach and analytical experiments undertaken in this study are summarized in Figure 3. We first examined the behavior of formaldehyde-containing adducts in the acidic DNPH solution to verify the factual significance of the theory asserted by Jensen (Jensen et al., 2015), suggesting a pseudo-irreversible conversion of hemiacetal to acetal (1,1-geminal diether) induced by a unidirectional shift in hemiacetal/acetal equilibrium. The latter phenomenon occurs, according to Jensen et al., under a synergic effect arisen from the low-pH environment and the high abundance of PG and Gly (containing 2 and 3 hydroxyl moieties, respectively) in the reaction condition.

Little is known with respect to the formation of acetals in aerosol. Additional puffing experiments were also conducted to verify the possibility of the formation of formaldehyde-acetal adducts during the aerosolization process.

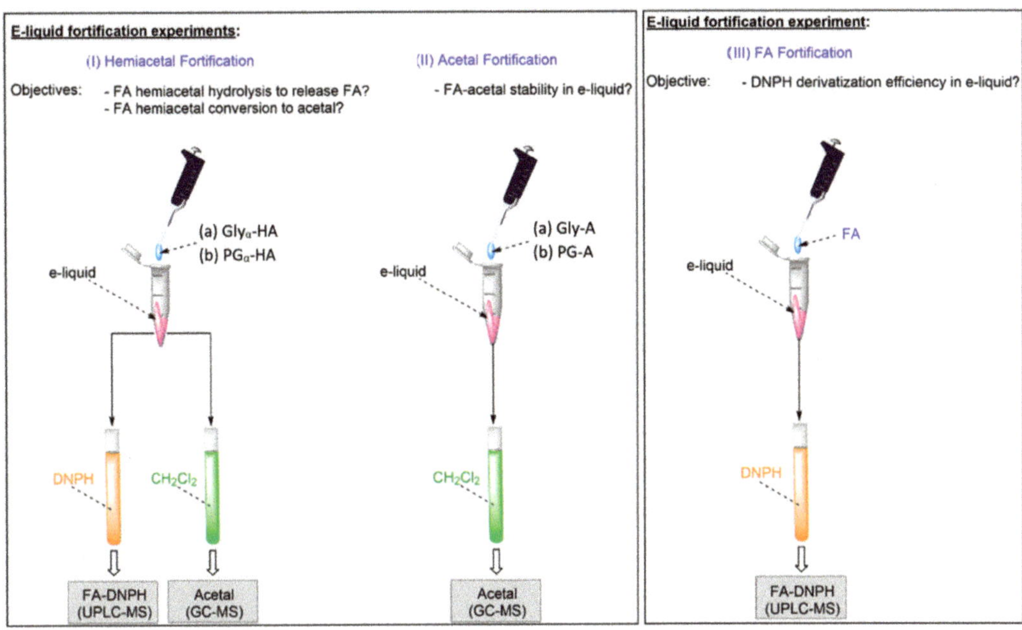

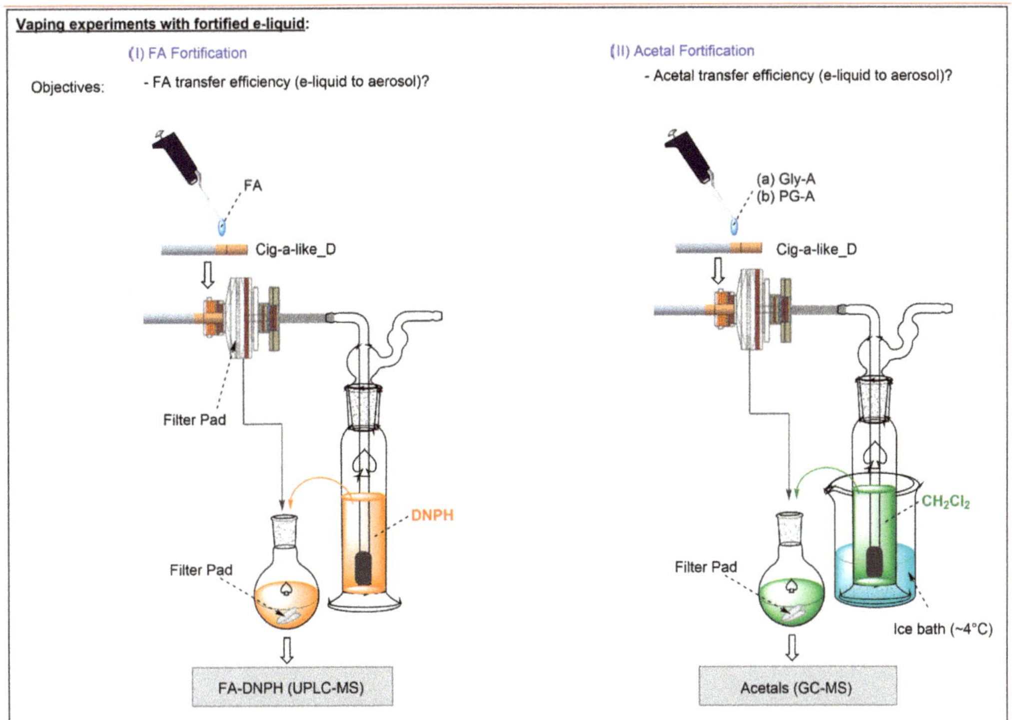

Figure 3. Summary of investigatory approach and analytical experiments.

3. Results and Discussion

3.1. Hemiacetal Behavior in Acidic DNPH Environment

3.1.1. Investigating Potential Intramolecular Cyclization for FA-Hemiacetal to Acetals

To determine whether hemiacetal adducts undergo hydrolysis in the acidic DNPH solution, 20 mg of PGα-HA and Glyα-HA was added into two separate 20 mL aliquots of DNPH derivatization solution. The fortified mixtures were shaken for 30 s and further diluted with additional DNPH solution, resulting in hemiacetal concentrations of 5.86 µg/mL (Glyα-HA) and 3.19 µg/mL (PGα-HA). The fortified mixtures were then treated according to the procedure described earlier for e-liquid samples (Materials and Methods section), and their FA-acetal adducts (PG-A and Gly-A) were quantified by GC–MS, as described in the Analytical Method subsection. The formation of acetal adducts was deemed "confirmed" by comparing the retention times and mass spectral data to the corresponding commercially available material.

Figure 4 illustrates chromatographic traces for acetal molecular species acquired in fortified mixtures and acetal standard solutions (approximately 5 µg/mL). No acetal adducts were detected in fortified samples, indicating that formaldehyde-hemiacetal adducts did not convert to their respective acetals in the studied reaction environment (i.e., acidic DNPH). These results contradict Jensen's theory [36] of the unidirectional shift in hemiacetal/acetal equilibrium to form acetal adducts in the acidic DNPH environment.

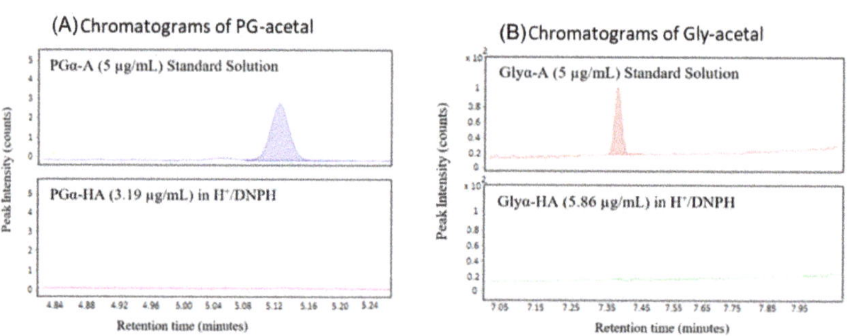

Figure 4. Comparison of chromatographic traces for PG-A (**A**) and Gly-A (**B**): DNPH fortified with PGα-HA and Glyα-HA (bottom) vs. standard solutions of FA-acetal adducts (top). FA-acetal adducts were analyzed by GC–MS, as described in the Analytical Method subsection.

3.1.2. Investigating Potential Hydrolysis of FA-Hemiacetal Adducts to Release Formaldehyde

Additional experiments were conducted to verify whether FA-containing hemiacetals can undergo hydrolysis to release FA in the acidic DNPH solution (Scheme 2). Known amounts of PGα-HA and Glyα-HA were added to the DNPH derivatization solution to yield concentrations at 3.19 µg/mL (PGα-HA) and 5.86 µg/mL (Glyα-HA). In the event FA-hemiacetal hydrolysis occurs, the released FA is assumed to be readily derivatized to generate FA-DNPH-hydrazone, which is quantifiable by UPLC–MS. The theoretical (expected) and measured formaldehyde concentrations in the acidic DNPH are reported in Table 2.

Table 2. Average (n = 3) conversion percentage of (Glyα-HA and PGα-HA) adducts to formaldehyde in acidic DNPH environment.

FA-Hemiacetal Adduct	[FA] Expected (µg/mL)	Average (n = 3) [FA] Measured (µg/mL)	% Hydrolysis of FA-HA Adducts in H⁺/DNPH
PGα-HA	0.93	0.96	103
Glyα-HA	1.44	1.50	104

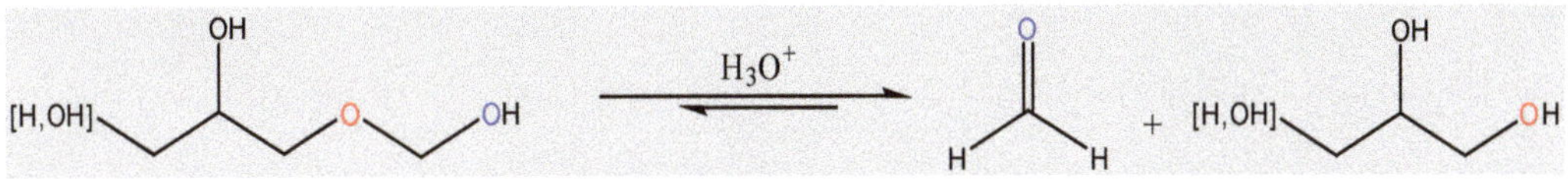

Scheme 2. FA-hemiacetal hydrolysis (colors are used to visualize the distribution of oxygen atoms between the reaction products).

The conversion rates (% hydrolysis) of 103–104% reported in Table 2 demonstrate that both hemiacetal adducts, i.e., Glyα-HA and PGα-HA, undergo complete hydrolysis in the acidic environment and release formaldehyde. The latter readily reacts with DNPH reagent (present in large excess) to form the corresponding hydrazone (FA-DNPH). These results are in agreement with Knorr's report that the FA-DNPH derivative yields measured in the aerosol extract cover both free formaldehyde, as well as formaldehyde from PGα-and Glyα-HA that may be present in the solution [42].

3.2. Acetal Reactivity and Formation Experiments

3.2.1. Investigating Hydrolysis of Cyclic Formaldehyde-Acetal Adducts (Gly-A and PG-A) Acidic DNPH Environment

The release of formaldehyde from cyclic Gly-A and PG-A adducts requires two consecutive acid-catalyzed hydrolytic reactions involving the formation of an intermediate hemiacetal (Gly-HA and PG-HA, respectively). This hypothesis was investigated by fortifying a reference e-liquid formula (15% water, 2.5% NBW in a 50/50 mixture of PG and Gly) with known amounts of PG-A and Gly-A adducts. The formation of the intermediate hemiacetal adducts in H^+/DNPH was investigated by measuring the FA-DNPH hydrazone formed between FA (released by complete hydrolysis of hemiacetal) and the DNPH reagent (UPLC–MS analysis). FA-DNPH hydrazone was not detected in the DNPH extract solution, suggesting that the intermediate FA-hemiacetal was either not formed or formed and readily released FA in the H^+/DNPH environment. This finding allows us to demonstrate that the hydrolysis of cyclic FA-acetal does not occur in the acidified DNPH environment.

3.2.2. Evaluation of Formaldehyde-Acetals (Gly-A and PG-A) Formation as a By-Product of the Aerosolization Process

Puffing experiments were conducted on both cig-a-like and pod-type products listed in Table 1, to verify the possibility of the formation of formaldehyde-acetal adducts (Gly-A and PG-A) during the aerosolization process. Immediately after aerosol collection, the filter pad was extracted in the impinger solution, and acetal adducts were quantified by GC–MS. Figures 5 and 6 summarize averaged acetal yields measured in e-liquids and aerosols for each product.

Gly-A Adduct: Figure 5 shows that, except for the cig-a-like CCE4 exhibiting relatively high Gly-A levels in both e-liquid and aerosol (~70 µg/g), the Gly-A levels in all other cig-a-like e-liquids and aerosol emissions ranged between 0.5 and 6 µg/g. With respect to cig-a-like products, we recorded an excellent correlation (linear regression, $R^2 = 0.999$) between Gly-A content in the e-liquids vs. its yield in the corresponding aerosol (Figure 5, inset plot). The Gly-A levels in aerosol were similar to those in the corresponding e-liquid. This observation led us to conclude that the presence of Gly-A in the cig-a-like device aerosol occurs predominantly from the transfer of the adduct from e-liquids into the corresponding aerosols, as opposed to the adduct being formed by an acetalization reaction taking place during the aerosolization process.

With respect to pod category products (Figure 6), one of the three devices (i.e., PGE11) showed a significant increase in Gly-A yield in aerosol (over 2.5 fold increase as compared to the e-liquid), indicating that Gly-A is also formed as an aerosolization by-product for this specific pod product.

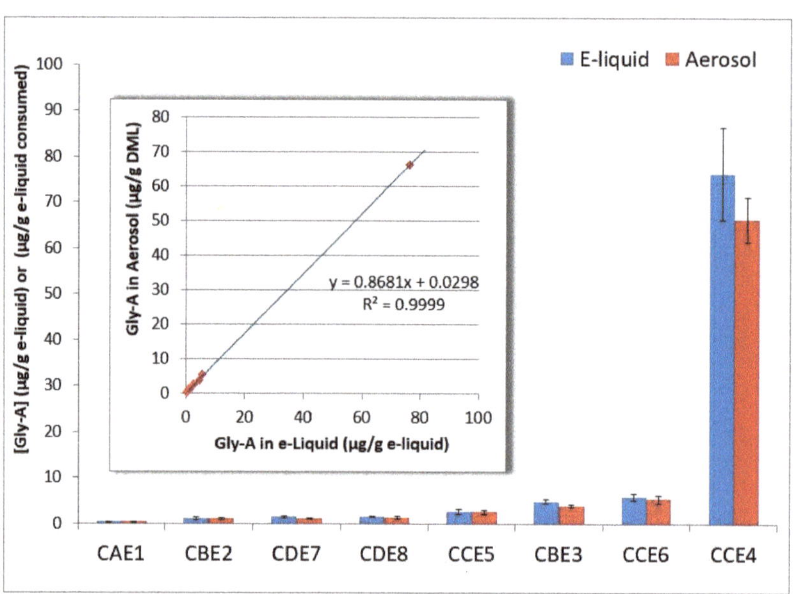

Figure 5. Gly-A average concentrations (n = 4) measured in e-liquid (μg/g e-liquid) and aerosol emission (μg/g e-liquid consumed) cig-a-like device category. Inset plot: e-liquid content vs. aerosol emissions.

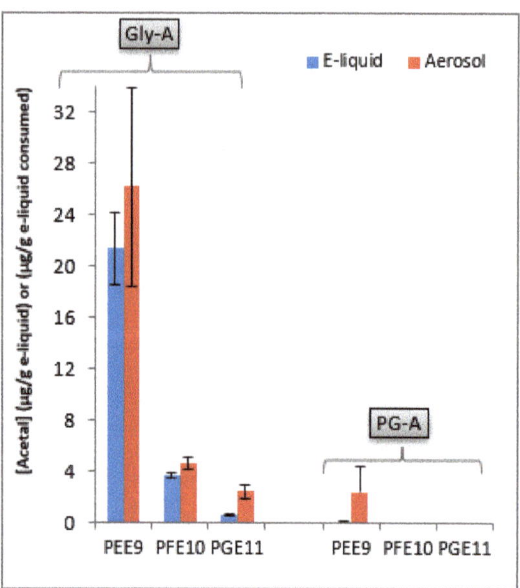

Figure 6. Gly-A and PG-A average concentrations (n = 4) measured in e-liquid (μg/g of e-liquid) and e-cigarette aerosol (μg/g of e-liquid consumed) Pod device category: Gly-A significantly increased in PGE11 aerosol while PG-A was only detected in PEE9. Error bars represent the standard deviation of each dataset.

PG-A Adduct: The PG-A was not detected in e-liquid or aerosol emission of any of the cig-a-like products investigated. With respect to pod category products, low μg/g levels were detected in one of the three tested pod products, i.e., PEE9 (Figure 6). The formation of PG-A in E9 e-liquid might be due to an acetalization reaction occurring in the e-liquid that may have flavor added. A recent study published in Nicotine & Tobacco Research [43] reported that acetalization reactions could occur between PG hydroxyl moieties and flavor aldehydes (i.e., benzaldehyde, cinnamaldehyde, citral, ethylvanillin, and vanillin) to form aldehyde acetals in chemically reactive e-liquids. With respect to puffing experiments, the PG-A yield in aerosols generated from the PE device exhibited a significant increase (-18 times), compared to its measured content in the E9-flavored e-liquid (Figure 6). The increase in the amount of PG-A adduct in the aerosol may be attributed to an acetalization reaction taking place during the aerosolization process of the E9-flavored e-liquid in the PE device. To evaluate and compare the reciprocal influences of e-liquid composition and e-cig design (emission profile) on the acetalization reaction, a series of acetal-fortification experiments were conducted using an unflavored reference e-liquid that are discussed in the next section.

3.3. Investigating Formation of PG-Acetal in PEE9 Aerosol: Formation during Aerosolization vs. Transfer from e-Liquid to Aerosol

To confirm the hypotheses put forward with regard to the acetal formation pathway (i.e., formation during aerosolization or transfer from e-liquid to aerosol), a series of fortification experiments were conducted in which known amounts of PG- and Gly-acetal adducts were fortified (separate experiments) into the reference formulation (15% water, 2.5% NBW in a 50/50 mixture of PG and Gly). The fortified e-liquids were loaded into empty PE cartridges. Aerosol collection and analytical procedures used for the quantification of acetal adducts in fortified e-liquids and their yields in aerosol emissions are described in Figure 3. The fortification amounts added to the unflavored reference e-liquid were such as to ensure that acetal levels (if formed) are above the method limit of quantification (0.8 μg/g of e-liquid or e-liquid consumed).

The results of quantitative analysis for acetal adducts (Gly-A and PG-A) are summarized and presented as the average of four replicate observations (Figure 7). To evaluate and compare acetal levels in the e-liquid and aerosol, their detected quantities are expressed in μg/g of e-liquid and μg/g of vaporized e-liquid, respectively. The Gly-A concentrations in e-liquid and aerosol phases are not statistically different. Conversely, PG-A concentration was augmented, on average, from 5.9 (e-liquid) to 9.3 (aerosol) μg/g, corresponding to an approximately 60% increase in PG-A during aerosolization. The increased amount of PG-A adduct in unflavored e-liquid aerosol (+60%) is markedly lower as compared to the increase (18 times) observed for the E9-flavored e-liquid aerosol (Figure 6). This observation led us to conclude that the formation of the PG acetal adduct (PG-A) in aerosol is predominantly driven by the flavor composition in the e-liquid E9 as opposed to the design of the PE device.

3.4. Evaluation of the Efficiency of DNPH Derivatization Method

To investigate the method accuracy for the quantification of FA in e-liquid and aerosol, two commercially available CDE7 (nonflavored) and CDE8 (flavored) cartridges were used. Prior to puffing, e-liquid contents (E7 and E8) were removed from 11 cartridges, combined for each sample type, aliquoted, and then fortified with known amounts of formaldehyde. The formaldehyde concentration in fortified e-liquid was at -20 μg/g of e-liquid. The unfortified (background level of formaldehyde in the matrix) and fortified e-liquid samples were loaded into empty CD cartridges. Aerosol collection and analyses of formaldehyde levels in e-liquid (prior to puffing) and e-cigarette aerosol were conducted as described earlier. Table 3 shows the method accuracy calculated from these fortification experiments when the method was applied to e-liquid and aerosol. The excellent formaldehyde recovery values of 97.1–105.5% reported in Table 3 indicate that the formaldehyde derivatization by DNPH is the predominant reaction under study conditions; therefore, the method is fit for the quantification of formaldehyde in e-cigarette products.

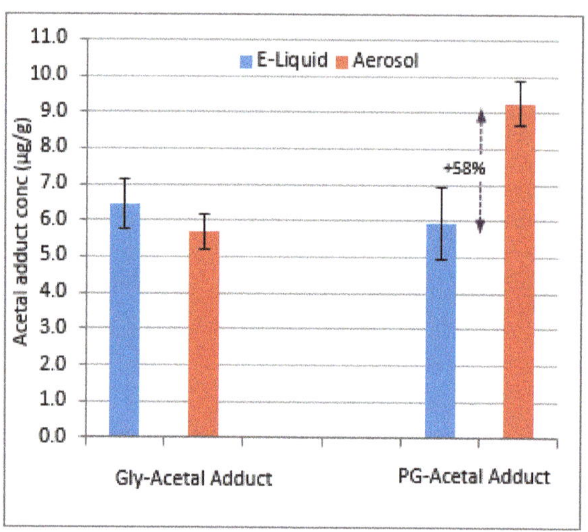

Figure 7. Comparison of average amounts (n = 4) of acetal adducts measured in fortified reference e-liquid (15% water, 2.5% NBW in a 50/50 mixture of PG and Gly) and aerosol emission (CD device).

Table 3. Method accuracy (fortification experiments, n = 4): %recovery values of formaldehyde in e-liquid and aerosol emission (CDE7 and CDE8) of 50 puffs.

e-Liquid		Unfortified Sample Concentration (μg/g)	Fortified Sample Concentration (μg/g)	Fortified Concentration (μg/g)	%Recovery (%)
CDE 7	Average	2.72	23.50	19.78	105.1
	SD	0.032	2.13		10.8
	%RSD	1.2	9.1		10
CDE 8	Average	14.35	33.90	19.92	98.1
	SD	0.12	0.27		1.3
	%RSD	0.84	0.80		1.4

Aerosol		Unfortified Sample Concentration (μg/g e-Liquid Consumed)	Fortified Sample Concentration (μg/g e-Liquid Consumed)	Fortified Concentration (μg/g)	%Recovery (%)
CDE 7	Average	19.14	38.45	19.78	97.6
	SD	2.07	2.17		11.0
	%RSD	11	5.6		11.2
CDE 8	Average	22.39	41.75	19.92	97.1
	SD	0.67	0.51		2.6
	%RSD	3.0	1.2		2.6

All concurring reactions/equilibria between participating reactants in the acidified environment are summarized in Figure 8. DNPH (0.6 mmoles) is in large excess as compared to the formaldehyde (0.04 mmoles, assuming an averaged FA emission rate of 25 μg/puff, 50 puffs collected). The reaction media is in a state of equilibrium governed by Le Chatelier's principle. The latter stipulates that a system in a state of equilibrium counteracts any perturbation by reaching a new equilibrium state. The consumption of formaldehyde by DNPH in the media is readily compensated by a shift in formaldehyde-hemiacetal hydrolysis (to release formaldehyde).

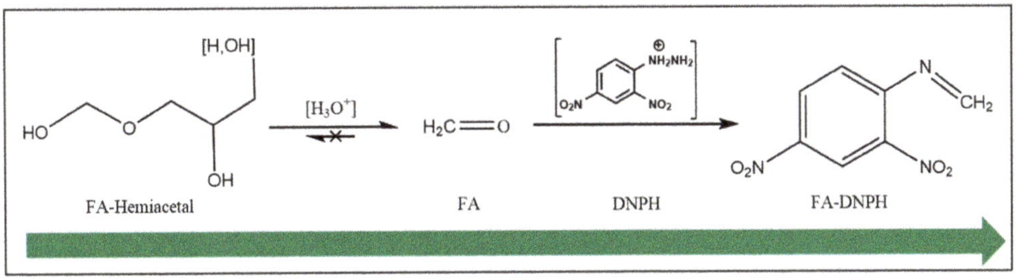

Figure 8. Concurring chain reactions between participating reactants in H+/DNPH environment.

4. Conclusions

The goal of this study was to provide an objective assessment of the DNPH method performance with respect to formaldehyde quantification in e-cigarette emissions. Our findings are in contradiction with a publication by the Jensen group (Jensen et al., 2015), which suggested that formaldehyde levels in ENDS products were underreported because formaldehyde may react with e-liquid excipients (PG and Gly) in the aerosol to form hemiacetals, which, in turn, form cyclic acetals in the acidic DNPH trapping solution.

The results from our investigations, focused on the behavior of formaldehyde-containing hemiacetal adducts in the acidic DNPH solution, clearly demonstrated that these compounds undergo a complete hydrolysis in the acidic environment to release formaldehyde, which is then derivatized by DNPH to form formaldehyde-hydrazone (FA-DNPH). Conversely, acetals of PG and Gly added to the DNPH trapping solution would not hydrolyze to form the hydrazone.

Our results from machine-generated aerosols showed that the aerosol generated from formaldehyde-fortified e-liquids provided quantitative recovery of formaldehyde in the aerosol, suggesting that if any hemiacetal was formed in the aerosol, it would readily hydrolyze to free formaldehyde in the acidic DNPH trapping solution. We believe that the presence of derivatization agent (DNPH) at a large excess in the acidic solution exerts a major role on hemiacetal/formaldehyde equilibrium: the hemiacetal/formaldehyde equilibrium shifts from the hemiacetal to the formaldehyde due to a complete and rapid consumption of free formaldehyde by DNPH.

We also demonstrated that acetal adducts fortified into e-liquids are almost completely transferred (−90%) to the aerosol during aerosolization in both device categories. Additionally, we observed that in the case of one of the tested pod devices (PE), the PG-acetal adduct can also be formed via an acetalization reaction during the aerosolization process.

We believe that our evidence-based analytic approach provides an objective assessment of DNPH method performance. The results of this study demonstrate that the measured FA-DNPH yields in the aerosol of the e-cigarettes account for all unreacted formaldehyde and formaldehyde–hemiacetal adducts and, therefore, the DNPH derivatization method allows for an accurate measurement of formaldehyde amounts in e-cigarette liquids and aerosols.

Author Contributions: Conceptualization, Y.B.P., K.A.W., W.P.G. and M.S.M.; formal analysis, X.C.J., R.M.B. and K.C.A.; investigation, M.S. and X.C.J.; methodology, M.S.M. and X.C.J.; project administration, X.C.J.; resources, R.M.B. and K.C.A.; supervision, K.A.W.; visualization, X.C.J. and M.S.; writing—original draft, M.S.; writing—review and editing, M.S., M.S.M., X.C.J., W.P.G. and Y.B.P. All authors have read and agreed to the published version of the manuscript.

Funding: This research received no external funding.

Institutional Review Board Statement: Not applicable.

Informed Consent Statement: Not applicable.

Acknowledgments: The authors would like to express their gratitude to Jennifer H. Smith, Jason W. Flora, Fadi Aldeek, Cynthia Cecil, and Robin Brownhill for their excellent comments and suggestions during review of this manuscript.

Conflicts of Interest: The authors declare no conflict of interest.

Abbreviations

CFP, Cambridge Filter Pad; CORESTA, Centre de Coopération pour les Recherches Scientifiques Relatives au Tabac (Centre for Scientific Research Relative to Tobacco); CRM, CORESTA recommended method; DNPH, 2,4-dinitrophenlhydrazine; ENDS, electronic nicotine delivery system; FA, formaldehyde; FA-DNPH, formaldehyde-DNPH hydrazone; FDA, U.S. Food and Drug Administration; GC–MS, gas chromatography–mass spectrometry; Gly, glycerin; Gly-A, glycerin-acetal adduct; Glyα-HA, glycerin-hemiacetal adduct; HPHC, harmful and potentially harmful constituent; HPLC-DAD, high-performance liquid chromatography-diode array detector; ICH, international conference on harmonization; ISO, international organization for standardization; LC–MS/MS, liquid chromatography–tandem mass spectrometry; NBW, nicotine by weight; PG, propylene glycol; PGα-HA, propylene glycol-hemiacetal adduct; PG-A, propylene glycol-acetal adduct; PFBHA, O-(2,3,4,5,6-pentafluorobenzyl)hydroxylamine; SIM, single ion monitoring; SPME, solid-phase microextraction; UPLC–MS/MS, ultra-performance liquid chromatography–tandem mass spectrometry; UV, ultraviolet; WHO, World Health Organization.

References

1. IARC. *Monographs on the Identification of Carcinogenic Hazards to Humans*; IARC: Lyon, France, 2012. Available online: https://monographs.iarc.who.int/list-of-classifications (accessed on 10 September 2021).
2. Chris, J. *Formaldehyde Resins Used in Industry, Manufacturing, and Construction*; Bright Hub Engineering: Troy, NY, USA, 2010. Available online: https://www.brighthubengineering.com/manufacturing-technology/88542-formaldehyde-resins-as-engineering-materials/ (accessed on 10 September 2021).
3. IARC. *Monographs on the Evaluation of Carcinogenic Risks to Humans: Formaldehyde, 2-Butoxyethanol and 1-tert-Butoxypropan-2-ol*; International Agency for Research on Cancer: Lyon, France, 2006.
4. Kaden, D.A.; Mandin, C.; Nielsen, G.D.; Wolkoff, P. *WHO Guidelines for Indoor Air Quality: Selected Pollutants*; World Health Organization: Geneva, Switzerland, 2010. Available online: https://www.ncbi.nlm.nih.gov/books/NBK138711/ (accessed on 10 September 2021).
5. Regulations Amending the Tobacco Reporting Regulations: SOR/2019-64. *Can. Gaz.* **2019**, *153 Pt II*, 6. Available online: https://canadagazette.gc.ca/rp-pr/p2/2019/2019-03-20/html/sor-dors64-eng.html (accessed on 10 September 2021).
6. Wright, C. Standardized methods for the regulation of cigarette-smoke constituents. *TrAC Trends Anal. Chem.* **2015**, *66*, 118–127. [CrossRef]
7. FDA. *Reporting Harmful and Potentially Harmful Constituents in Tobacco Products and Tobacco Smoke Under Section 904(a)(3) of the Federal Food, Drug, and Cosmetic Act. Draft Guidance for Industry*; U.S. Department of Health and Human Services, Food and Drug Administration: Rockville, MD, USA, 2012.
8. FDA. Harmful and potentially harmful constituents in tobacco products and tobacco smoke; established list. *Fed. Regist.* **2012**, *77*, 20034–20037.
9. FDA. *Premarket Tobacco Product Applications for Electronic Nicotine Delivery Systems: Guidance for Industry*; U.S. Department of Health and Human Services, Food and Drug Administration, Center for Tobacco Products: Rockville, MD, USA, 2019.
10. Reilly, S.M.; Goel, R.; Trushin, N.; Elias, R.J.; Foulds, J.; Muscat, J.; Liao, J.; Richie, J.P. Brand variation in oxidant production in mainstream cigarette smoke: Carbonyls and free radicals. *Food Chem. Toxicol.* **2017**, *106*, 147–154. [CrossRef]
11. Counts, M.; Morton, M.; Laffoon, S.; Cox, R.; Lipowicz, P. Smoke composition and predicting relationships for international commercial cigarettes smoked with three machine-smoking conditions. *Regul. Toxicol. Pharmacol.* **2005**, *41*, 185–227. [CrossRef]
12. Roemer, E.; Stabbert, R.; Rustemeier, K.; Veltel, D.; Meisgen, T.; Reininghaus, W.; Carchman, R.; Gaworski, C.; Podraza, K. Chemical composition, cytotoxicity and mutagenicity of smoke from US commercial and reference cigarettes smoked under two sets of machine smoking conditions. *Toxicology* **2004**, *195*, 31–52. [CrossRef]
13. Cheng, T. Chemical evaluation of electronic cigarettes. *Tob. Control.* **2014**, *23* (Suppl S2), ii11–ii17. [CrossRef]
14. Gillman, I.; Kistler, K.; Stewart, E.; Paolantonio, A. Effect of variable power levels on the yield of total aerosol mass and formation of aldehydes in e-cigarette aerosols. *Regul. Toxicol. Pharmacol.* **2016**, *75*, 58–65. [CrossRef]
15. Khlystov, A.; Samburova, V. Flavoring Compounds Dominate Toxic Aldehyde Production during E-Cigarette Vaping. *Environ. Sci. Technol.* **2016**, *50*, 13080–13085. [CrossRef] [PubMed]
16. Kosmider, L.; Sobczak, A.; Fik, M.; Knysak, J.; Zaciera, M.; Kurek, J.; Goniewicz, M. Carbonyl compounds in electronic cigarette vapors: Effects of nicotine solvent and battery output voltage. *Nicotine Tob. Res.* **2014**, *16*, 1319–1326. [CrossRef]

17. Laino, T.; Tuma, C.; Curioni, A.; Jochnowitz, E.; Stolz, S. A revisited picture of the mechanism of glycerol dehydration. *J. Phys. Chem. A* **2011**, *115*, 3592–3595. [CrossRef]
18. Ohta, K.; Uchiyama, S.; Inaba, Y.; Nakagome, H.; Kunugita, N. Determination of carbonyl compounds generated from the electronic cigarette using coupled silica cartridges impregnated with hydroquinone and 2,4-dinitrophenylhydrazine. *Anal. Sci.* **2011**, *60*, 791–797.
19. Paine, J.B.; Pithawalla, Y.B.; Naworal, J.D.; Thomas, C.E. Carbohydrate pyrolysis mechanisms from isotopic labeling: Part 1: The pyrolysis of glycerin: Discovery of competing fragmentation mechanisms affording acetaldehyde and formaldehyde and the implications for carbohydrate pyrolysis. *J. Anal. Appl. Pyrolysis* **2007**, *80*, 297–311. [CrossRef]
20. Uchiyama, S.; Ohta, K.; Inaba, Y.; Kunugita, N. Determination of carbonyl compounds generated from the e-cigarette using coupled silica cartridges impregnated with hydroquinone and 2,4-dinitrophenylhydrazine, followed by high-performance liquid chromatography. *Anal. Sci.* **2013**, *29*, 1219–1222. [CrossRef] [PubMed]
21. Wang, P.; Chen, W.; Liao, J.; Matsuo, T.; Ito, K.; Fowles, J.; Shusterman, D.; Mendell, M.; Kumagai, K. A Device-Independent Evaluation of Carbonyl Emissions from Heated Electronic Cigarette Solvents. *PLoS ONE* **2017**, *12*, e0169811. [CrossRef] [PubMed]
22. Goniewicz, M.L.; Knysak, J.; Gawron, M.; Kosmider, L.; Sobczak, A.; Kurek, J.; Prokopowicz, A.; Jabłońska-Czapla, M.; Rosik-Dulewska, C.; Havel, C.; et al. Levels of selected carcinogens and toxicants in vapour from electronic cigarettes. *Tob. Control.* **2013**, *23*, 133–139. [CrossRef] [PubMed]
23. Margham, J.; McAdam, K.; Forster, M.; Liu, C.; Wright, C.; Mariner, D.; Proctor, C. Chemical Composition of Aerosol from an E-Cigarette: A Quantitative Comparison with Cigarette Smoke. *Chem. Res. Toxicol.* **2016**, *29*, 1662–1678. [CrossRef]
24. Beauval, N.; Verriele, M.; Garat, A.; Fronval, I.; Dusautoir, R.; Anthérieu, S.; Garçon, G.; Lo-Guidice, J.M.; Allorge, D.; Locoge, N. Influence of puffing conditions on the carbonyl composition of e-cigarette aerosols. *Int. J. Hyg. Environ. Health* **2019**, *222*, 136–146. [CrossRef]
25. Farsalinos, K.E.; Voudris, V.; Poulas, K. E-cigarettes generate high levels of aldehydes only in 'dry puff' conditions. *Addiction* **2015**, *110*, 1352–1356. [CrossRef]
26. Flora, J.W.; Meruva, N.; Huang, C.B.; Wilkinson, C.T.; Ballentine, R.; Smith, D.C.; Werley, M.S.; McKinney, W.J. Characterization of potential impurities and degradation products in electronic cigarette formulations and aerosols. *Regul. Toxicol. Pharmacol.* **2016**, *74*, 1–11. [CrossRef]
27. Geiss, O.; Bianchi, I.; Barrero-Moreno, J. Correlation of volatile carbonyl yields emitted by e-cigarettes with the temperature of the heating coil and the perceived sensorial quality of the generated vapours. *Int. J. Hyg. Environ. Health* **2016**, *219*, 268–277. [CrossRef]
28. Farsalinos, K.E.; Voudris, V.; Spyrou, A.; Poulas, K. E-cigarettes emit very high formaldehyde levels only in conditions that are aversive to users: A replication study under verified realistic use conditions. *Food Chem. Toxicol.* **2017**, *109*, 90–94. [CrossRef]
29. CORESTA. Recommended Method No. 74: Determination of Selected Carbonyls in Mainstream Cigarette Smoke by HPLC 2019. Available online: www.coresta.org/sites/default/files/technical_documents/main/CRM_74-Aug2019_0.pdf (accessed on 10 September 2021).
30. Health Canada. Official Method T-104; Determination of Selected CarBonyls in Mainstream Tobacco Smoke 1999. Available online: https://www.healthycanadians.gc.ca/en/open-information/tobacco/t100/carbonyl (accessed on 1 May 2019).
31. ISO 21160:2018. Cigarettes—Determination of Selected Carbonyls in the Mainstream Smoke of Cigarettes—Method Using High Performance Liquid Chromatography 2018. Available online: https://www.iso.org/standard/69993.html (accessed on 1 May 2019).
32. Bao, M.; Joza, P.J.; Masters, A.; Rickert, W.S. Analysis of selected carbonyl compounds in tobacco samples by using pentafluorobenzylhydroxylamine derivatization and gas chromatography-mass spectrometry. *Beitr. Tabakforsch. Int.* **2014**, *26*, 86–97. [CrossRef]
33. Ogunwale, M.A.; Li, M.; Raju, M.V.R.; Chen, Y.; Nantz, M.H.; Conklin, D.J.; Fu, X.-A. Aldehyde Detection in Electronic Cigarette Aerosols. *ACS Omega* **2017**, *2*, 1207–1214. [CrossRef] [PubMed]
34. Sala, C.; Medana, C.; Pellegrino, R.; Aigotti, R.; Bello, F.D.; Bianchi, G.; Davoli, E. Dynamic measurement of newly formed carbonyl compounds in vapors from electronic cigarettes. *Eur. J. Mass Spectrom.* **2017**, *23*, 64–69. [CrossRef] [PubMed]
35. Salamanca, J.C.; Munhenzva, I.; Escobedo, J.O.; Jensen, R.P.; Shaw, A.; Campbell, R.; Luo, W.; Peyton, D.H.; Strongin, R.M. Formaldehyde Hemiacetal Sampling, Recovery, and Quantification from Electronic Cigarette Aerosols. *Sci. Rep.* **2017**, *7*, 11044. [CrossRef]
36. Jensen, R.P.; Luo, W.; Pankow, J.F.; Strongin, R.M.; Peyton, D.H. Hidden Formaldehyde in E-Cigarette Aerosols. *N. Engl. J. Med.* **2015**, *372*, 392–394. [CrossRef]
37. Bates, C.D.; Farsalinos, K.E. Research letter on e-cigarette cancer risk was so misleading it should be retracted. *Addiction* **2015**, *110*, 1686–1687. [CrossRef] [PubMed]
38. Nitzkin, J.L.; Farsalinos, K.; Siegel, M. More on hidden formaldehyde in e-cigarette aerosols. *N. Engl. J. Med.* **2015**, *372*, 1575.
39. Sleiman, M.; Logue, J.M.; Montesinos, N.; Russell, M.L.; Litter, M.I.; Gundel, L.A.; Destaillats, H. Emissions from Electronic Cigarettes: Key Parameters Affecting the Release of Harmful Chemicals. *Environ. Sci. Technol.* **2016**, *50*, 9644–9651. [CrossRef]
40. Flora, J.W.; Wilkinson, C.T.; Wilkinson, J.W.; Lipowicz, P.J.; Skapars, J.A.; Anderson, A.; Miller, J.H. Method for the Determination of Carbonyl Compounds in E-Cigarette Aerosols. *J. Chromatogr. Sci.* **2017**, *55*, 142–148. [CrossRef] [PubMed]

41. International Conference on Harmonisation of Technical Requirements for Registration of Pharmaceuticals for Human Use. ICH Harmonised Tripartite Guideline: Validation of Analytical Procedures: Text and Methodology Q2(R1) 2005. Available online: https://database.ich.org/sites/default/files/Q2_R1__Guideline.pdf (accessed on 1 May 2019).
42. Knorr, A.; Gautier, L.; Debrick, A.; Tekeste, E.; Buchholz, C.; Almstetter, M.; Arndt, D.; Bentley, M. *Formaldehyde-Glycerol Hemiacetal—Absence of "Hidden" Formaldehyde in THS 2.2. Aerosols*; Presented at 3D Cell; DECHEMA: Zurich, Switzerland, 2012.
43. Erythropel, H.C.; Jabba, S.V.; Dewinter, T.M.; Mendizabal, M.; Anastas, P.T.; Jordt, S.E.; Zimmerman, J.B. Formation of flavorant–propylene Glycol Adducts With Novel Toxicological Properties in Chemically Unstable E-Cigarette Liquids. *Nicotine Tob. Res.* **2019**, *21*, 1248–1258. [CrossRef] [PubMed]

Article

Targeted Characterization of the Chemical Composition of JUUL Systems Aerosol and Comparison with 3R4F Reference Cigarettes and IQOS Heat Sticks

Xin Chen [†], Patrick C. Bailey, Clarissa Yang, Bryant Hiraki, Michael J. Oldham and I. Gene Gillman *

JUUL Labs Inc., 1000 F St. NW, Washington, DC 20004, USA; xinchen_2000@yahoo.com (X.C.); patrick.bailey@juul.com (P.C.B.); clarissa.yang@juul.com (C.Y.); bryant.hiraki@juul.com (B.H.); michael.oldham@juul.com (M.J.O.)

* Correspondence: gene.gillman@juul.com

† Current Address: Turn Biotechnologies, 319 N. Bernardo Ave., Mountain View, CA 94043, USA.

Abstract: Aerosol constituent yields have been reported from a wide range of electronic nicotine delivery systems. No comprehensive study has been published on the aerosol constituents generated from the JUUL system. Targeted analyses of 53 aerosol constituents from the four JUUL products currently on the US market (Virginia Tobacco and Menthol flavored e-liquids in both 5.0% and 3.0% nicotine concentration by weight) was performed using non-intense and intense puffing regimens. All measurements were conducted by an ISO 17025 accredited contract research organization. JUUL product aerosol constituents were compared to published values for the 3R4F research cigarette and IQOS Regular and Menthol heated tobacco products. Across the four JUUL products and two puffing regimes, only 10/53 analytes were quantifiable, including only two carbonyls (known propylene glycol or glycerol degradants). The remaining analytes were primary ingredients, nicotine degradants and water. Average analyte reductions (excluding primary ingredients and water) for all four JUUL system aerosols tested were greater than 98% lower than 3R4F mainstream smoke, and greater than 88% lower than IQOS aerosol. In summary, chemical characterization and evaluation of JUUL product aerosols demonstrates a significant reduction in toxicants when compared to mainstream cigarette smoke from 3R4F reference cigarettes or aerosols from IQOS-heated tobacco products.

Keywords: JUUL; aerosol; HPHC; GC-MS

Citation: Chen, X.; Bailey, P.C.; Yang, C.; Hiraki, B.; Oldham, M.J.; Gillman, I.G. Targeted Characterization of the Chemical Composition of JUUL Systems Aerosol and Comparison with 3R4F Reference Cigarettes and IQOS Heat Sticks. *Separations* **2021**, *8*, 168. https://doi.org/10.3390/separations8100168

Academic Editors: Wojciech Piekoszewski and Beatriz Albero

Received: 31 July 2021
Accepted: 24 September 2021
Published: 3 October 2021

Publisher's Note: MDPI stays neutral with regard to jurisdictional claims in published maps and institutional affiliations.

1. Introduction

A number of compounds in tobacco smoke have been recognized by the US Food and Drug Administration (FDA) as harmful and potentially harmful constituents (HPHCs) [1] and the Agency has required the reporting of these toxicant levels in mainstream cigarette smoke [2]. These compounds have toxicities relevant to a number of tobacco related diseases such as cancer, and cardiovascular and respiratory diseases [3]. In contrast to combustible cigarettes, electronic nicotine delivery system/s (ENDS) are designed to deliver nicotine without combustion [4]. The devices themselves consist of a battery, a heating element (most often a coil), and a reservoir for storing e-liquid. A number of studies have investigated whether this is reflected in a reduced toxicant profile of ENDS aerosol and concluded that compounds such as carbonyls [5–12], tobacco-specific nitrosamines [13–15], polycyclic aromatic hydrocarbons (PAH) [14,16,17], volatile organic compounds [8,14,18], and others [19,20] are significantly reduced in comparison to the levels in mainstream cigarette smoke. Correspondingly, a number of scientific bodies have concluded that completely substituting ENDS products for combustible cigarettes may reduce a smoker's exposure to toxicants, including carcinogens [21,22].

However, some publications have also reported the production of elevated levels of carbonyl analytes and other HPHCs in ENDS aerosol. This has included reports of

73

formaldehyde levels in ENDS aerosols that have approached, or even exceeded, levels in mainstream cigarette smoke [9,10,23–25]. Kosmider and colleagues analyzed the aerosol of a second generation ENDS device, the eGo-3, and reported that formaldehyde content was dependent on e-liquid formulation and device power [26]. When the variable power battery was raised to its highest setting, carbonyl content approached that of mainstream cigarette smoke. Jensen et al. reported similar findings with an unnamed tank system ENDS [27]. The authors reported the detection of no formaldehyde at low voltage, but formaldehyde content exceeding that of mainstream cigarette smoke at high voltage.

More recently, formaldehyde, acetaldehyde, and acrolein levels have been shown to be highly dependent on the power levels of the ENDS device [24,28]. A recent study examining the effects of high-power settings illustrated that for every 5 watts of increase above manufacturer recommended power settings, carbonyl generation increased by up to 20 times [28]. An additional study found that higher than manufacturer recommended power settings not only increase the carbonyl content of ENDS aerosol, but are aversive to the sensory experience of the user [25]. Previous work has established that higher coil temperatures lead to increased thermal degradation of the e-liquid, leading to the excess production of a range of HPHCs [29]. The wide range of HPHC levels reported in ENDS indicates that device performance, including power delivered to the coil and coil temperature, have a substantial impact on the production of HPHCs in the ENDS aerosols.

There are a large variety of ENDS products with a vast array of device characteristics on the market. Over the last 14 years, ENDS products have undergone several significant changes in design. The fourth generation of ENDS device, the "pod mod" or simply "pod", generally consist of a small, rechargeable battery and an insertable pod which comes pre-filled with e-liquid. Most e-liquids consist of humectants such as propylene glycol (PG) and vegetable glycerol (VG), as well as nicotine and flavorants. Device characteristics are known to affect the HPHC emissions of ENDS devices [30]. One important device characteristic is the ability to limit coil temperature based on changes to the resistance of the coil during heating. The JUUL system includes tightly controlled temperature regulation of the coil, with the goal of minimizing potential heat degradation by-products and HPHCs in the resulting aerosols [31]. Unlike prior work [31,32], the targeted analysis described in this paper combined with the non-targeted analysis described in our companion paper provide a more complete analysis of aerosol constituents in the JUUL device [33].

The JUUL system is a pre-filled, closed-system, cartridge-based ENDS product that consists of two main components: a JUUL device and a pre-filled JUULpod. The primary e-liquid ingredients in a JUULpod are nicotine, propylene glycol (PG), vegetable glycerin (VG), benzoic acid, and flavorants. Although no single study has comprehensively characterized the aerosol chemistry of the JUUL system, it has been reported to produce lower levels of carbonyls, free radicals, and CO than mainstream cigarette smoke and other tested ENDS products [32]. The JUUL device itself is designed to operate below combustion temperatures [34]. Talih and colleagues have reported reduced carbonyl emissions in the JUUL system and attributed the reduced carbonyl levels, versus other ENDS products, to the lower coil temperature of the JUUL system [31,35]. The primary objective of this research was to quantify an expanded set of aerosol constituents generated from the four US products (Virginia Tobacco 5% [VT5], Virginia Tobacco 3% [VT3], Menthol 5% [Me5], and Menthol 3% [Me3]). The secondary objective of this research was to compare the measured aerosol constituent levels of the four JUUL products to the reported literature values from the mainstream cigarette smoke of the 3R4F reference cigarette and the aerosol from the IQOS Tobacco Heating System.

2. Materials and Methods

2.1. Test Items

JUULpods used in testing were commercial products stored in their original commercially marketed packaging and tested within two months of pod filling. Aerosol constituents chosen for targeted analysis were based on the FDA Final Guidance on PM-

TAs for ENDS (Group I) and the FDA Draft Guidance on PMTAs for ENDS (Group II) (Table 1) [36,37]. Many of the aerosol constituents listed in Table 1 are also classified by the FDA as HPHCs [1,38]. The Virginia Tobacco and Menthol JUULpods used in this study contained 3.0% and 5.0% nicotine on a mass-to-mass basis and 35 and 59 mg/mL on a mass-to-volume basis.

Table 1. Aerosol Constituents Included in the Study.

Aerosol Constituent Names		
Group I		**Group II**
1-Butanol	Glycerol	1-Aminonaphthalene
β-Nicotyrine	Glycidol *	1,3-Butadiene
Acetaldehyde	Isoamyl Acetate	2-Aminonaphthalene
Acetyl Propionyl	Isobutyl Acetate	4-Aminobiphenyl
Acrolein	Isobutyraldehyde	Ammonia
Acrylonitrile	Lead	Anabasine
Benzene	Menthol	Anatabine
Benzoic Acid	Methyl Acetate	Benzo(a)pyrene
Benzyl Acetate	n-Butyraldehyde	Carbon Monoxide
Cadmium	Nickel	Gold
Chromium	Nicotine	Isoprene
Copper	Nicotine-N-Oxide	Myosmine
Cotinine	NNK	Nornicotine
Crotonaldehyde	NNN	
Diacetyl	Propionic Acid	
Diethylene Glycol	Propylene Glycol	
Ethyl Acetate	Propylene Oxide	
Ethyl Acetoacetate	Toluene	
Ethylene Glycol	Water	
Formaldehyde		
Furfural		

*: Method unavailable at the time of the study.

2.2. Generation and Collection of Aerosol

Group 1 aerosol constituent analysis was conducted by Labstat International ULC (Labstat; Kitchener, Ontario, Canada). Group II aerosol constituent analysis were performed by Enthalpy Analytical LLC (800 Capitola Drive, Suite 1, Durham, NC, 27713 and 1470 East Parham Road, Richmond, VA, 23228). Both Labstat International ULC and Enthalpy Analytical were International Organization for Standardization (ISO) 17025 accredited at the time of this study. All analytical methods were validated for the analysis of ENDS aerosol according to ICH guidance Q2 (R1) with the exception of gold and carbon monoxide [39]. Method validations included an assessment of accuracy, precision, repeatability, intermediate precision, specificity, detection limit, quantitation limit, linearity, and recovery from the trapping systems. All method validations were reviewed by an independent accreditation body as part of the ISO 17025 accreditation process. Carbon monoxide was determined following ISO 8454 and gold was determined by ICP-MS and method performance was verified for accuracy, detection limit, quantitation limit, and linearity [40]. A summary of the analytical methods used in this study are presented in Supplemental Table S1.

JUUL devices used in the testing were commercial products. Prior to aerosol collection, the JUULpod was attached to a fully charged device. Devices were replaced every 50 puffs during aerosol collection of Group I aerosol constituents. All aerosol collections were performed on linear puffing machines and the JUUL System was inserted into a custom pad holder containing a glass fiber filter pad to trap non-volatile compounds during aerosol collection. Depending on the test method, an impinger containing a trapping solvent may have been used in conjunction with, or instead of the glass fiber pad

(Supplemental Table S1). The JUUL device was oriented at a 45° angle to gravity with the battery end downward.

For Group I aerosol constituents, a total of ten replicate measurements were performed from each of three lots of each JUUL flavor/concentration (n = 30) for each puffing regimen. Data for Group II aerosol constituents was based upon the FDA Draft Guidance on PMTAs for ENDS and utilized 10 replicate measurements of one batch of each of the four JUUL products. Aerosol samples were collected under two puffing regimes: non-intense (NI) and intense. NI puffing collection for Group I analytes was conducted consistent with ISO 20768 (55 mL puff volume over 3 s with one puff every 30 s). For Group II analytes, which were collected prior to the existence of ISO 20768, a 70 mL puff volume was used as opposed to 55 mL [41]. Currently, there is no standardized topography for intense ENDS puffing conditions. The intense puffing regime employed here used a 110 mL puff volume over a 6 s puff duration (maximum possible with the JUUL device) with one puff every 30 s.

At present, there is no standardized method for the collection of ENDS product aerosols. The yield of aerosol mass and selected aerosol constituents has been shown to vary across the life of a device with aerosol mass decreasing and analyte levels increasing as the e-liquid is depleted [42,43]. To determine the impact of total puff count on aerosol mass and aerosol constituent yield, Group I aerosol constituents were analyzed over three 50-puff collections: one at the beginning (first 50 puffs), one in the middle (45–50% of the total device mass loss (DML)), and one at the end (85–90% of total DML). For Group II aerosol constituents, only the beginning 50-puff segment was analyzed.

2.3. Measurement of Aerosol Constituents

All contract research organization (CRO) aerosol constituent measurement methods were validated and included in their scope of accreditation when the analyses were performed. Methods are summarized in Supplemental Table S1. Air blank samples were collected and analyzed together with the JUULpod aerosol samples for each method.

2.4. Estimated Values

The majority of the JUUL systems aerosol constituents were below the limit of detection (BLOD) or below limit of quantification (BLOQ). To facilitate comparison, when a constituent in JUUL system aerosol was BLOD, its level was computed as half of reported LOD; when the constituent was BLOQ, the level was considered as the average of reported LOD and LOQ [12]. A potential limitation of using estimated values are instances where an estimated value is larger than a quantified value, as it is not possible to determine if the difference is a reduction or an increase.

2.5. Background Subtraction

To mitigate the impact of environmental background during aerosol collection, which can lead to false-positives and/or overestimation of results, laboratory background control (air blank) measurements were performed [12]. Blank background subtraction was applied to aerosol sample datasets when reporting results for ammonia, chromium, formaldehyde, and lead. The background subtraction approach was applied separately to the respective puff segment group (i.e., puffs 1–50 segment analyte measurements were compared against puffs 1–50 segment blank values). When evaluating specific puff segment collections, no numerical imputed values were applied when a not different from blank (NDFB) measurement was determined. In cases where results for the air blank average value were below the level of detection (BLOD) or below the limit of quantitation (BLOQ), no action was taken. In cases where results for the air blank average value and the analyte average value are above the LOQ, and the average air blank value was greater than or equal to the average analyte value, the analyte value was reported as NDFB. In cases where results for the air blank value were non-zero, a statistical analysis was performed using Student's t-test (unpaired, nonparametric, 2-tailed) to establish if the sample and the blank results were significantly different ($p < 0.05$). If there was no statistically significant difference, the analyte result was reported as NDFB. If the background and sample results were statistically different, the difference between the sample mean and the background mean was computed (sample mean minus blank mean).

2.6. Comparators Testing and Value Sources

The primary comparator used in this work was the 3R4F Kentucky Reference Cigarette (University of Kentucky, USA). IQOS 2.2 Regular and IQOS 2.2 Menthol (mIQOS) heated tobacco products were used as secondary comparators.

Aerosol constituent values for mainstream 3R4F reference cigarette smoke were obtained from peer-reviewed literature for both ISO Non-Intense [44] and ISO Intense [45] smoking regimes. Values for the majority of mainstream smoke constituents were taken from Jaccard et al., 2019 [46]; Acetyl propionyl and diacetyl from Moldoveneau et al., 2017 [47]; chromium, lead and nickel from Pappas et al., 2014 [48]. IQOS and mIQOS aerosol constituent values were obtained from the literature and public sources [49,50] for ISO Intense smoking regime. ISO Non-Intense values were not available for IQOS aerosols. Source data, by constituent, is given in Supplemental Tables S4–S15.

2.7. Data Processing for Comparison between Test Systems

Comparison of aerosol constituent values from ENDS products to the values in mainstream cigarette smoke is a non-trivial task. Cigarettes may be consumed in 10–15 puffs when smoked with an intense puffing regime [12]. In contrast, a VT5 JUULpod puffed under our intense regimen lasts for slightly more than 300 puffs. Dividing collection values by the number of puffs provides a means to compare the products on a per puff basis, but this may not be representative of real-world usage, as nicotine product users are known to modify their topographies in order to titrate their nicotine intake to desired levels [51–53]. With this in mind, in addition to per puff, aerosol constituent values were normalized to nicotine content by dividing the targeted aerosol constituent value by the measured value of nicotine from the same study. This provides aerosol constituent intake on a per nicotine basis. In general, the highest reported JUUL aerosol constituent levels across the three puff blocks (beginning, middle, and end) were used as the basis for comparison to 3R4F and IQOS. The highest aerosol constituent level across the three puff blocks is designated in bold text in Table 2. When values were BLOD for all three puff blocks, the value from the first 50 puffs was used for comparison.

Accordingly, we normalized reported aerosol constituent values for IQOS and the 3R4F reference cigarette to nicotine as well. This allowed a more direct comparison to determine whether analytes in JUUL aerosols were reduced vs. 3R4F and/or IQOS aerosols. JUUL aerosol constituents generated using NI and intense regimes were compared directly to equivalent regimes in the comparator products.

The following equation was used to calculate the percent difference between JUUL and comparator products.

% difference = ((JUUL aerosol constituent level normalized by nicotine)/(Comparator constituent level normalized by nicotine) − 1) × 100%

The % Difference value was regarded as Not Comparable (NC) if (1) both JUUL and comparator were BLOD/BLOQ, (2) the measured JUUL value was NDFB, or (3) the measured JUUL value was BLOD or BLOQ and the comparator value was quantifiable, but lower than the method LOD/LOQ.

Table 2. Quantifiable analytes in the aerosol of VT5 (per puff).

| | | | | VT5 Per Puff Values | | | | | |
| | | | | NI | | | Intense | | |
HPHC or Chemical	Units	LOD	LOQ	Beginning	Middle	End	Beginning	Middle	End
Benzoic Acid	mg	9.59×10^{-05}	3.20×10^{-04}	4.14×10^{-02}	6.37×10^{-02}	**7.81×10^{-02}**	**7.82×10^{-02}**	1.37×10^{-01}	2.88×10^{-02}
Cotinine	µg	1.87×10^{-03}	6.25×10^{-03}	$\leq 4.06 \times 10^{-03}$	$\leq 4.06 \times 10^{-03}$	$\leq 4.06 \times 10^{-03}$	$\leq 4.06 \times 10^{-03}$	**6.57×10^{-03}**	$\leq 4.06 \times 10^{-03}$
Formaldehyde	µg	5.49×10^{-03}	1.83×10^{-02}	2.72×10^{-02}	**8.33×10^{-02}**	7.77×10^{-02}	2.60×10^{-02}	**6.03×10^{-02}**	5.43×10^{-02}
Glycerol	mg	1.44×10^{-03}	4.80×10^{-03}	7.37×10^{-01}	$1.06 \times 10^{+00}$	**$1.24 \times 10^{+00}$**	$1.35 \times 10^{+00}$	**$2.12 \times 10^{+00}$**	4.53×10^{-01}
Nicotine	mg	1.35×10^{-04}	4.49×10^{-04}	5.43×10^{-02}	8.49×10^{-02}	**1.05×10^{-01}**	9.99×10^{-02}	**1.72×10^{-01}**	3.77×10^{-02}
Nornicotine	*µg*	*4.76×10^{-03}*	*1.59×10^{-02}*	***2.42×10^{-02}***	*NT*	*NT*	***2.42×10^{-02}***	*NT*	*NT*
Propylene Glycol	mg	2.40×10^{-04}	8.01×10^{-04}	2.40×10^{-01}	3.96×10^{-01}	**5.32×10^{-01}**	4.60×10^{-01}	**8.30×10^{-01}**	2.17×10^{-01}
Water	mg	3.83×10^{-03}	1.28×10^{-02}	1.11×10^{-01}	2.21×10^{-01}	**2.82×10^{-01}**	2.57×10^{-01}	**4.31×10^{-01}**	1.10×10^{-01}
VT3 Per Puff Values									
Acetaldehyde	µg	9.95×10^{-03}	3.32×10^{-02}	$\leq 2.16 \times 10^{-02}$	**3.47×10^{-02}**	$\leq 2.16 \times 10^{-02}$	$\leq 2.16 \times 10^{-02}$	**4.35×10^{-02}**	$\leq 2.16 \times 10^{-02}$
Benzoic Acid	mg	9.59×10^{-05}	3.20×10^{-04}	2.61×10^{-02}	4.33×10^{-02}	**4.79×10^{-02}**	4.23×10^{-02}	**8.41×10^{-02}**	3.23×10^{-02}
Formaldehyde	µg	5.49×10^{-03}	1.83×10^{-02}	2.89×10^{-02}	**7.24×10^{-02}**	6.17×10^{-02}	1.19×10^{-02}	**8.72×10^{-02}**	4.68×10^{-02}
Glycerol	mg	1.44×10^{-03}	4.80×10^{-03}	7.26×10^{-01}	**$1.23 \times 10^{+00}$**	$1.19 \times 10^{+00}$	$1.29 \times 10^{+00}$	**$2.40 \times 10^{+00}$**	9.26×10^{-01}
Nicotine	mg	1.35×10^{-04}	4.49×10^{-04}	2.93×10^{-02}	5.35×10^{-02}	**5.67×10^{-02}**	5.28×10^{-02}	**1.05×10^{-01}**	4.35×10^{-02}
Nornicotine	*µg*	*4.76×10^{-03}*	*1.59×10^{-02}*	***1.66×10^{-02}***	*NT*	*NT*	***1.66×10^{-02}***	*NT*	*NT*
Propylene Glycol	mg	2.40×10^{-04}	8.01×10^{-04}	2.31×10^{-01}	4.47×10^{-01}	**5.01×10^{-01}**	4.48×10^{-01}	**9.19×10^{-01}**	4.07×10^{-01}
Water	mg	3.83×10^{-03}	1.28×10^{-02}	1.75×10^{-01}	**2.75×10^{-01}**	2.37×10^{-01}	3.00×10^{-01}	**4.57×10^{-01}**	2.49×10^{-01}
Me5 Per Puff Values									
β-Nicotyrine	µg	3.81×10^{-03}	1.27×10^{-02}	$\leq 1.90 \times 10^{-03}$	$\leq 8.25 \times 10^{-03}$	$\leq 8.25 \times 10^{-03}$	$\leq 1.90 \times 10^{-03}$	**1.33×10^{-02}**	1.31×10^{-02}
Acetaldehyde	µg	9.95×10^{-03}	3.32×10^{-02}	$\leq 2.16 \times 10^{-02}$	$\leq 2.16 \times 10^{-02}$	$\leq 2.16 \times 10^{-02}$	$\leq 2.16 \times 10^{-02}$	$\leq 2.16 \times 10^{-02}$	**3.57×10^{-02}**
Benzoic Acid	mg	9.59×10^{-05}	3.20×10^{-04}	3.24×10^{-02}	**7.30×10^{-02}**	6.67×10^{-02}	5.43×10^{-02}	**1.17×10^{-01}**	1.16×10^{-01}
Ethylene Glycol	mg	5.05×10^{-05}	1.68×10^{-04}	$\leq 1.09 \times 10^{-04}$	$\leq 1.09 \times 10^{-04}$	$\leq 1.09 \times 10^{-04}$	$\leq 1.09 \times 10^{-04}$	$\leq 1.09 \times 10^{-04}$	**1.78×10^{-04}**
Formaldehyde	µg	5.49×10^{-03}	1.83×10^{-02}	1.19×10^{-02}	**3.94×10^{-02}**	3.44×10^{-02}	1.19×10^{-02}	5.33×10^{-02}	**7.56×10^{-02}**
Glycerol	mg	1.44×10^{-03}	4.80×10^{-03}	6.00×10^{-01}	**$1.23 \times 10^{+00}$**	$1.09 \times 10^{+00}$	9.92×10^{-01}	$1.96 \times 10^{+00}$	**$1.99 \times 10^{+00}$**
Menthol	mg	2.44×10^{-04}	8.14×10^{-04}	7.31×10^{-03}	**1.81×10^{-02}**	1.67×10^{-02}	1.20×10^{-02}	2.85×10^{-02}	**2.90×10^{-02}**
Nicotine	mg	1.35×10^{-04}	4.49×10^{-04}	4.03×10^{-02}	**8.93×10^{-02}**	8.24×10^{-02}	6.92×10^{-02}	**1.56×10^{-01}**	1.55×10^{-01}
Nornicotine	*µg*	*4.76×10^{-03}*	*1.59×10^{-02}*	***1.99×10^{-05}***	*NT*	*NT*	***1.99×10^{-05}***	*NT*	*NT*
Propylene Glycol	mg	2.40×10^{-04}	8.01×10^{-04}	1.78×10^{-01}	4.16×10^{-01}	**4.24×10^{-01}**	3.50×10^{-01}	7.58×10^{-01}	**7.83×10^{-01}**
Water	mg	3.83×10^{-03}	1.28×10^{-02}	1.79×10^{-01}	**2.52×10^{-01}**	2.46×10^{-01}	2.42×10^{-01}	3.60×10^{-01}	**3.84×10^{-01}**
Me3 Per Puff Values									
Acetaldehyde	µg	9.95×10^{-03}	3.32×10^{-02}	$\leq 2.16 \times 10^{-02}$	$\leq 2.16 \times 10^{-02}$	$\leq 2.16 \times 10^{-02}$	$\leq 2.16 \times 10^{-02}$	$\leq 2.16 \times 10^{-02}$	**3.56×10^{-02}**
Benzoic Acid	mg	9.59×10^{-05}	3.20×10^{-04}	2.07×10^{-02}	**4.10×10^{-02}**	4.08×10^{-02}	3.76×10^{-02}	7.24×10^{-02}	**8.57×10^{-02}**
Cotinine	µg	1.87×10^{-03}	6.25×10^{-03}	9.37×10^{-04}	$\leq 4.06 \times 10^{-03}$	$\leq 4.06 \times 10^{-03}$	$\leq 4.06 \times 10^{-03}$	$\leq 4.06 \times 10^{-03}$	$\leq 4.06 \times 10^{-03}$
Ethylene Glycol	mg	5.05×10^{-05}	1.68×10^{-04}	**$\leq 1.09 \times 10^{-04}$**	$\leq 1.09 \times 10^{-04}$	$\leq 1.09 \times 10^{-04}$	2.20×10^{-04}	1.09×10^{-04}	1.09×10^{-04}
Formaldehyde	µg	5.49×10^{-03}	1.83×10^{-02}	NDFB	1.85×10^{-02}	2.19×10^{-02}	NDFB	3.78×10^{-02}	**4.47×10^{-02}**
Glycerol	mg	1.44×10^{-03}	4.80×10^{-03}	6.78×10^{-01}	**$1.22 \times 10^{+00}$**	$1.16 \times 10^{+00}$	$1.16 \times 10^{+00}$	$2.14 \times 10^{+00}$	**$2.46 \times 10^{+00}$**
Menthol	mg	2.44×10^{-04}	8.14×10^{-04}	8.07×10^{-03}	1.74×10^{-02}	**1.79×10^{-02}**	1.32×10^{-02}	3.08×10^{-02}	**3.44×10^{-02}**
Nicotine	mg	1.35×10^{-04}	4.49×10^{-04}	2.55×10^{-02}	4.98×10^{-02}	**5.04×10^{-02}**	4.56×10^{-02}	9.38×10^{-02}	**1.07×10^{-01}**
Propylene Glycol	mg	2.40×10^{-04}	8.01×10^{-04}	2.03×10^{-01}	4.20×10^{-01}	**4.55×10^{-01}**	3.97×10^{-01}	8.20×10^{-01}	**9.92×10^{-01}**
Water	mg	3.83×10^{-03}	1.28×10^{-02}	1.47×10^{-01}	**2.51×10^{-01}**	2.38×10^{-01}	2.81×10^{-01}	3.87×10^{-01}	**4.95×10^{-01}**

Note: Values preceded with "$\leq$" are based on estimated values. Analytes in italics were tested under a single modified puffing regimen: 70mL puff, 3 s duration, 30 s interval. Bolded values are highest of the three puff blocks.

3. Results

A total of 45/53 aerosol constituents were BLOD/BLOQ in JUUL VT5/VT3, and 43/53 aerosol constituents were BLOD/BLOQ in JUUL Me5/Me3. Quantifiable aerosol constituents (mean value > LOQ in at least one puff block in one puffing regime) in JUUL products are presented on a per puff basis in Table 2. Method LODs and LOQs for all 53 aerosol constituents are presented in Supplemental Tables S2 and S3.

The values in bold are highest of the three puff blocks. Only 10 of the 53 aerosol constituents were quantifiable in any of the JUUL systems aerosols. All of these aerosol constituents were in Group I and were collected at the beginning, middle and end of pod life (Table 2). Aerosol mass was found to increase after collection of the beginning puff block. Yields of formaldehyde and acetaldehyde were highest in the middle and end puff blocks with 50% of the highest values in the end puff block.

Aerosol constituent values for each JUUL product using both puffing regimes were then normalized to nicotine and compared to 3R4F and IQOS (Supplemental Tables S4–S15). Across all flavors and nicotine concentrations, of the aerosol constituents which could be compared, all were reduced in JUUL aerosols compared to 3R4F mainstream cigarette smoke, excepting water and the primary ingredients PG and VG (Table 4). Water was reduced in all JUUL aerosols, except in the NI regimens of VT3 and Me3. Notably, 22/25 of the aerosol constituent reductions were based on estimated values for JUUL aerosol constituents as they were either BLOD or BLOQ in JUUL products. Only formaldehyde ($\downarrow \geq$96% to $\geq$99%) and acetaldehyde ($\downarrow \geq$99%) were quantifiable (Table 3). More comprehensive results for each JUUL flavor and concentration are outlined below.

Table 3. Summary of Aerosol Constituent Comparisons between the JUUL System and 3R4F Cigarette Smoke.

Aerosol Constituents Higher in all JUUL System Aerosols vs. 3R4F Smoke	
PG (RT)	**VG (RT)**
Aerosol Constituents Lower in all JUUL Aerosols vs. 3R4F Smoke	
1-aminonaphthalene (CA) [$\downarrow \geq$99%] *	CO (RDT) [$\downarrow \geq$98% to $\geq$99%] *
1,3-butadiene (CA, RT, RDT) [$\downarrow \geq$99%] *	Crotonaldehyde (CA) [$\downarrow \geq$99%] *
2-aminonaphthalene (CA) [$\downarrow \geq$99%] *	Diacetyl (RT) [$\downarrow \geq$99%] *
4-aminobiphenyl (CA) [$\downarrow \geq$99%] *	**Formaldehyde (CA, RT) [$\downarrow \geq$96% to $\geq$99%]**
Acetaldehyde (CA, RT, AD) [$\downarrow \geq$99%]	Furfural (RT) [$\downarrow \geq$91% to $\geq$97%] *
Acetyl propionyl (RT) [$\downarrow \geq$99%] *	Isoprene (CA) [$\downarrow \geq$98% to $\geq$99%] *
Acrolein (RT, CT) [$\downarrow \geq$99%] *	Lead (CA, CT, RDT) [$\downarrow \geq$86% to $\downarrow = \geq$97%] *
Acrylonitrile (CA, RT) [$\downarrow \geq$99%] *	n-Butyraldehyde (RT) [$\downarrow \geq$99%] *
Ammonia (RT) [$\downarrow \geq$85% to $\geq$96%] *	NNK (CA) [$\downarrow \geq$99%] *
Benzene (CA, CT, RDT) [$\downarrow \geq$99%] *	NNN (CA) [$\downarrow \geq$99%] *
Benzo(a)pyrene (CA) [$\downarrow \geq$98% to $\geq$99%] *	Propylene oxide (CA, RT) [$\downarrow \geq$99%] *
Cadmium (CA, RT, RDT) [$\downarrow \geq$99%] *	Toluene * (RT, RDT) [$\downarrow \geq$99%] *

Note: Constituents with (*) were BLOD/Q in JUUL aerosols and reductions were based on estimated values. CA = Carcinogen. CT = Cardiovascular toxicant. RT = Respiratory Toxicant. AD = Addictive. RDT = Reproductive and developmental toxicant. $\downarrow$ = less than. Bolded reductions were based on quantifiable values.

With regard to IQOS, across all flavors and nicotine concentrations of JUUL products, of the aerosol constituents that were compared, only PG, VG, water ($\downarrow \geq$81% to $\geq$90%), formaldehyde ($\downarrow \geq$80% to $\geq$91%) and acetaldehyde ($\downarrow \geq$99%) were quantifiable. Twelve of the 15 aerosol constituent comparisons were based on estimated values. Seven aerosol constituents were lower in all JUUL aerosols vs. IQOS where comparisons could be made (e.g., one or more products were not compared with IQOS, but those which were compared were reduced), and cadmium, nickel, and chromium were BLOD/BLOQ/NDFB in both JUUL and IQOS products (Table 4). More comprehensive results for each JUUL flavor and concentration are outlined below.

Table 4. Summary of Constituent Comparisons between JUUL System and IQOS.

Aerosol Constituents Higher in JUUL System Aerosols vs. IQOS	
PG (RT)	**VG (RT)**
Aerosol Constituents Lower in all JUUL Aerosols vs. IQOS	
1-aminonaphthalene (CA) [↓≥66% to ≥86%] * **Acetaldehyde (CA, RT, AD) [↓≥99%]** Acrolein (RT, CT) [↓≥97%] * Acrylonitrile (CA, RT) [↓≥61% to ≥83%] * Benzene (CA, CT, RDT) [↓≥92% to ≥96%] * Benzo(a)pyrene (CA) [↓≥77% to ≥97%] * Crotonaldehyde (CA) [↓≥97%] * Diacetyl (RT) [↓≥99%] *	**Formaldehyde (CA, RT) [↓≥80% to ≥91%]** n-Butyraldehyde (RT) [↓≥99%] * NNK (CA) [↓≥96%] * NNN (CA) [↓≥99%] * Propylene oxide (CA, RT) [↓≥72% to ≥86%] * Toluene * (RT, RDT) [↓≥93% to ≥96%] * **Water [↓≥81% to≥90%]**
Aerosol Constituents Lower in all JUUL Aerosols vs. IQOS where comparisons could be made	
2-Aminonaphthalene (CA) [↓≥75% to ≥84%] * 1,3-butadiene (CA, RT, RDT) [↓≥64% to ≥80%] * Ammonia (RT) [↓≥83% to ≥94%] * CO (RDT) [↓≥68% to ≥80%] * Isoprene (CA) [↓≥98% to ≥99%] * Lead (CA, CT, RDT) [↓≥86% to ≥97%] *	Menthol [↓≥85% to ≥91%] *
Aerosol Constituents BLOD/BLOQ/NDFB in both JUUL and IQOS aerosols	
Cadmium Chromium	Nickel

Note: Constituents with (*) were BLOD/Q in JUUL aerosols and reductions were based on estimated values. CA = Carcinogen. CT = Cardiovascular toxicant. RT = Respiratory Toxicant. AD = Addictive. RDT = Reproductive and developmental toxicant. ↓ = less than. Bolded reductions were based on quantifiable values.

3.1. Virginia Tobacco

Table 5 and Supplemental Table S16 outline the comparisons of VT5 and VT3 aerosol constituents to 3R4F mainstream cigarette smoke and IQOS aerosol constituents.

Table 5. Comparison of Aerosol Constituents in VT5 aerosol generated using the intense puffing regime vs. comparator products.

Aerosol Constituent	Aerosol Constituent Normalized by Nicotine (mg/mg)					% Difference		
	VT5		3R4F		IQOS	vs. 3R4F		vs. IQOS
	NI	Intense	NI	Intense	Intense	NI	Intense	Intense
1-Aminonaphthalene (CA)	$\leq1.96 \times 10^{-08\,a}$	$\leq1.96 \times 10^{-08\,a}$	2.02×10^{-05}	1.46×10^{-05}	5.83×10^{-08}	↓≥99.90	↓≥99.87	↓≥66.39
1-Butanol	$\leq5.52 \times 10^{-04\,a}$	$\leq3.00 \times 10^{-04\,a}$	N/A	N/A	N/A	N/A	N/A	N/A
1,3-Butadiene (CA)	$\leq6.83 \times 10^{-05\,a}$	$\leq6.83 \times 10^{-05\,a}$	5.40×10^{-02}	5.03×10^{-02}	2.23×10^{-04}	↓≥99.87	↓≥99.86	↓≥69.34
2-Aminonaphthalene (CA)	$\leq8.53 \times 10^{-09\,a}$	$\leq8.53 \times 10^{-09\,a}$	1.25×10^{-05}	9.10×10^{-06}	3.48×10^{-08}	↓≥99.93	↓≥99.91	↓≥75.53
4-Aminobiphenyl (CA)	$\leq1.48 \times 10^{-08\,a}$	$\leq1.48 \times 10^{-08\,a}$	2.19×10^{-06}	1.96×10^{-06}	6.98×10^{-09}	↓≥99.32	↓≥99.24	NC
Acetaldehyde *(CA)*	$\leq3.97 \times 10^{-04\,b}$	$\leq2.16 \times 10^{-04\,b}$	8.50×10^{-01}	9.24×10^{-01}	1.66×10^{-01}	↓≥99.95	↓≥99.98	↓≥99.87
Acetyl Propionyl	$\leq3.23 \times 10^{-05\,a}$	$\leq1.76 \times 10^{-05\,a}$	4.21×10^{-02}	4.62×10^{-02}	N/A	↓≥99.92	↓≥99.96	N/A
Acrolein	$\leq1.92 \times 10^{-04\,a,b}$	$\leq1.17 \times 10^{-04\,a,b}$	7.58×10^{-02}	9.05×10^{-02}	8.56×10^{-03}	↓≥99.75	↓≥99.87	↓≥98.64
Acrylonitrile *(CA)*	$\leq5.89 \times 10^{-05\,a}$	$\leq3.20 \times 10^{-05\,a}$	7.45×10^{-03}	1.03×10^{-02}	1.95×10^{-04}	↓≥99.21	↓≥99.69	↓≥83.61
Ammonia	NDFB	NDFB	1.37×10^{-02}	1.71×10^{-02}	1.08×10^{-02}	NC	NC	NC
Anabasine	$\leq2.98 \times 10^{-04\,a}$	$\leq2.98 \times 10^{-04\,a}$	N/A	N/A	N/A	N/A	N/A	N/A
Anatabine	$\leq5.70 \times 10^{-05\,a}$	$\leq5.70 \times 10^{-05\,a}$	N/A	N/A	N/A	N/A	N/A	N/A
Benzene *(CA)*	$\leq3.14 \times 10^{-05\,a}$	$\leq1.71 \times 10^{-05\,a}$	4.79×10^{-02}	4.46×10^{-02}	4.92×10^{-04}	↓≥99.93	↓≥99.96	↓≥96.53

Table 5. *Cont.*

Aerosol Constituent	Aerosol Constituent Normalized by Nicotine (mg/mg)					% Difference		
	VT5		3R4F		IQOS	vs. 3R4F		vs. IQOS
	NI	Intense	NI	Intense	Intense	NI	Intense	Intense
Benzo(a)pyrene (CA)	$\leq 1.27 \times 10^{-07\,a}$	$\leq 1.27 \times 10^{-07\,a}$	8.97×10^{-06}	7.59×10^{-06}	5.71×10^{-07}	$\downarrow \geq 98.58$	$\downarrow \geq 98.32$	$\downarrow \geq 77.70$
Benzoic Acid	$\mathbf{7.45 \times 10^{-01}}$	$\mathbf{7.94 \times 10^{-01}}$	N/A	N/A	N/A	N/A	N/A	N/A
Benzyl Acetate	$\leq 5.52 \times 10^{-04\,a}$	$\leq 3.00 \times 10^{-04\,a}$	N/A	N/A	N/A	N/A	N/A	N/A
β-Nicotyrine	$\leq 9.72 \times 10^{-05\,a,b}$	$\leq 4.79 \times 10^{-05\,a,b}$	N/A	N/A	N/A	N/A	N/A	N/A
Cadmium *(CA)*	$\leq 9.66 \times 10^{-08\,a}$	$\leq 5.26 \times 10^{-08\,a}$	3.49×10^{-05}	4.68×10^{-05}	$\leq 2.65 \times 10^{-07}$	$\downarrow \geq 99.72$	$\downarrow \geq 99.89$	NC
Carbon Monoxide	$\leq 1.26 \times 10^{-01\,a}$	$\leq 1.26 \times 10^{-01\,a}$	$1.42 \times 10^{+01}$	$1.53 \times 10^{+01}$	4.02×10^{-01}	$\downarrow \geq 99.12$	$\downarrow \geq 99.18$	$\downarrow \geq 68.75$
Chromium	NDFB	NDFB	$\leq 1.25 \times 10^{-06}$	$\leq 6.84 \times 10^{-06}$	$\leq 4.17 \times 10^{-07}$	NC	NC	NC
Copper	$\leq 2.71 \times 10^{-06\,a}$	$\leq 6.57 \times 10^{-07\,a}$	N/A	N/A	N/A	N/A	N/A	N/A
Cotinine	$\leq 7.48 \times 10^{-05\,a}$	$\mathbf{3.81 \times 10^{-05}}$	N/A	N/A	N/A	N/A	N/A	N/A
Crotonaldehyde *(CA)*	$\leq 5.73 \times 10^{-05\,a}$	$\leq 3.12 \times 10^{-05\,a}$	1.41×10^{-02}	2.99×10^{-02}	3.14×10^{-03}	$\downarrow \geq 99.59$	$\downarrow \geq 99.90$	$\downarrow \geq 99.01$
Diacetyl	$\leq 1.60 \times 10^{-05\,a}$	$\leq 8.71 \times 10^{-06\,a}$	1.64×10^{-01}	1.69×10^{-01}	3.58×10^{-02}	$\downarrow \geq 99.99$	$\downarrow \geq 99.99$	$\downarrow \geq 99.98$
Diethylene Glycol	$\leq 2.21 \times 10^{-03\,a}$	$\leq 1.20 \times 10^{-03\,a}$	N/A	N/A	N/A	N/A	N/A	N/A
Ethyl Acetate	$\leq 5.52 \times 10^{-04\,a}$	$\leq 3.45 \times 10^{-03\,a,b}$	N/A	N/A	N/A	N/A	N/A	N/A
Ethyl Acetoacetate	$\leq 4.42 \times 10^{-05\,a}$	$\leq 2.40 \times 10^{-05\,a}$	N/A	N/A	N/A	N/A	N/A	N/A
Ethylene Glycol	$\leq 2.01 \times 10^{-03\,a}$	$\leq 1.09 \times 10^{-03\,b}$	N/A	N/A	N/A	N/A	N/A	N/A
Formaldehyde *(CA)*	$\mathbf{9.82 \times 10^{-04}}$	$\mathbf{3.50 \times 10^{-04}}$	3.52×10^{-02}	4.38×10^{-02}	4.19×10^{-03}	$\downarrow\mathbf{97.21}$	$\downarrow\mathbf{99.20}$	$\downarrow\mathbf{91.64}$
Furfural	$\leq 7.73 \times 10^{-04\,a}$	$\leq 4.20 \times 10^{-04\,a}$	1.84×10^{-02}	1.49×10^{-02}	2.41×10^{-02}	$\downarrow \geq 95.80$	$\downarrow \geq 97.18$	$\downarrow \geq 98.26$
Glycerol	$\mathbf{1.19 \times 10^{+01}}$	$\mathbf{1.23 \times 10^{+01}}$	$3.85 \times 10^{+00}$	$1.36 \times 10^{+00}$	$3.51 \times 10^{+00}$	$\mathbf{2.09 \times 10^{+02}}$	$\mathbf{8.08 \times 10^{+02}}$	$\mathbf{2.51 \times 10^{+02}}$
Glycidol *(CA)*	NR	NR	N/A	8.84×10^{-04}	4.43×10^{-03}	NC	NC	NC
Gold	$\leq 1.52 \times 10^{-07\,a}$	$\leq 1.52 \times 10^{-07\,a}$	N/A	N/A	N/A	N/A	N/A	N/A
Isoamyl Acetate	$\leq 8.83 \times 10^{-04\,a}$	$\leq 4.80 \times 10^{-04\,a}$	N/A	N/A	N/A	N/A	N/A	N/A
Isobutyl Acetate	$\leq 5.52 \times 10^{-04\,a}$	$\leq 3.00 \times 10^{-04\,a}$	N/A	N/A	N/A	N/A	N/A	N/A
Isobutyraldehyde	$\leq 1.52 \times 10^{-05\,a}$	$\leq 8.27 \times 10^{-06\,a}$	N/A	N/A	N/A	N/A	N/A	N/A
Isoprene (CA)	$\leq 9.72 \times 10^{-05\,a}$	$\leq 9.72 \times 10^{-05\,a}$	4.10×10^{-01}	4.02×10^{-01}	1.78×10^{-03}	$\downarrow \geq 99.98$	$\downarrow \geq 99.98$	$\downarrow \geq 94.54$
Lead	$\leq 4.34 \times 10^{-07\,a}$	$\leq 4.55 \times 10^{-07\,b}$	1.31×10^{-05}	1.79×10^{-05}	1.73×10^{-06}	$\downarrow \geq 96.69$	$\downarrow \geq 97.46$	$\downarrow \geq 73.65$
Menthol	$\leq 2.25 \times 10^{-03\,a}$	$\leq 1.22 \times 10^{-03\,a}$	$\leq 1.32 \times 10^{-05}$	$\leq 4.76 \times 10^{-06}$	3.42×10^{-04}	NC	NC	NC
Methyl Acetate	$\leq 6.62 \times 10^{-04\,a}$	$\leq 3.60 \times 10^{-04\,a}$	N/A	N/A	N/A	N/A	N/A	N/A
Myosmine	$\leq 4.62 \times 10^{-04\,a}$	$\leq 4.62 \times 10^{-04\,a}$	N/A	N/A	N/A	N/A	N/A	N/A
n-Butyraldehyde	$\leq 3.23 \times 10^{-05\,a}$	$\leq 1.76 \times 10^{-05\,a}$	4.05×10^{-02}	4.67×10^{-02}	1.98×10^{-02}	$\downarrow \geq 99.92$	$\downarrow \geq 99.96$	$\downarrow \geq 99.91$
Nickel	$\leq 8.37 \times 10^{-06\,a}$	$\leq 1.05 \times 10^{-06\,a}$	$\leq 5.41 \times 10^{-07}$	$\leq 7.41 \times 10^{-06}$	$\leq 4.17 \times 10^{-07}$	NC	NC	NC
Nicotine	$\mathbf{1.00 \times 10^{+00}}$	$\mathbf{1.00 \times 10^{+00}}$	$1.00 \times 10^{+00}$	$1.00 \times 10^{+00}$	$1.00 \times 10^{+00}$	N/A	N/A	N/A
Nicotine-N-Oxide	$\leq 1.60 \times 10^{-04\,a}$	$\leq 8.73 \times 10^{-05\,a}$	N/A	N/A	N/A	N/A	N/A	N/A
NNK *(CA)*	$\leq 1.39 \times 10^{-07\,a}$	$\leq 7.54 \times 10^{-08\,a}$	1.61×10^{-04}	1.48×10^{-04}	5.08×10^{-06}	$\downarrow \geq 99.91$	$\downarrow \geq 99.95$	$\downarrow \geq 98.52$
NNN *(CA)*	$\leq 9.06 \times 10^{-08\,a}$	$\leq 4.93 \times 10^{-08\,a}$	1.87×10^{-04}	1.69×10^{-04}	1.30×10^{-05}	$\downarrow \geq 99.95$	$\downarrow \geq 99.97$	$\downarrow \geq 99.62$
Nornicotine	$\mathbf{5.78 \times 10^{-04\,a}}$	$\mathbf{5.78 \times 10^{-04\,a}}$	N/A	N/A	N/A	N/A	N/A	N/A
Propionic Acid	$\leq 3.31 \times 10^{-04\,a}$	$\leq 1.80 \times 10^{-04\,a}$	N/A	N/A	N/A	N/A	N/A	N/A
Propylene Glycol	$\mathbf{5.07 \times 10^{+00}}$	$\mathbf{4.82 \times 10^{+00}}$	4.21×10^{-03}	1.36×10^{-02}	1.36×10^{-01}	$\mathbf{1.20 \times 10^{+05}}$	$\mathbf{3.52 \times 10^{+04}}$	$\mathbf{3.45 \times 10^{+03}}$
Propylene Oxide *(CA)*	$\leq 2.87 \times 10^{-05\,a}$	$\leq 1.56 \times 10^{-05\,a}$	4.23×10^{-01}	5.03×10^{-01}	1.12×10^{-04}	$\downarrow \geq 99.99$	$\downarrow \geq 100.00$	$\downarrow \geq 86.07$
Toluene	$\leq 1.13 \times 10^{-04\,a}$	$\leq 6.12 \times 10^{-05\,a}$	7.32×10^{-02}	7.69×10^{-02}	1.96×10^{-03}	$\downarrow \geq 99.85$	$\downarrow \geq 99.92$	$\downarrow \geq 96.88$
Water	$\mathbf{2.69 \times 10^{+00}}$	$\mathbf{2.50 \times 10^{+00}}$	$4.21 \times 10^{+00}$	$8.36 \times 10^{+00}$	$2.77 \times 10^{+01}$	$\downarrow 36.14$	$\downarrow 70.11$	$\downarrow 90.96$
						Average[c] $\downarrow\mathbf{99.31}$	**Average**[c] $\downarrow\mathbf{99.53}$	**Average**[c] $\downarrow\mathbf{88.72}$

BLOD = Below limit of quantitation; BLOQ = Below limit of quantitation; NDFB = Not different from background; NC = Not comparable; NR = Not reported; $\downarrow$ = less than. Note: Values preceded with "$\geq$" are based on estimated values. Values in bold were quantifiable. CA = carcinogen. Analytes in italics were tested under a single modified puffing regimen: 70 mL puff, 3 s duration, 30 s interval. [a] analyte was BLOD for all three 50 puff blocks. [b] analyte was BLOQ for all three 50 puff blocks. [a,b] analyte was either BLOD or BLOQ for all three 50 puff blocks. [c] excluding PG, VG, and water.

Although not all aerosol constituents examined in this targeted analysis were reported in the literature for the mainstream smoke of the 3R4F reference cigarette, of the compounds unavailable for comparison, only benzoic acid, cotinine (VT5-intense), and nornicotine were detectable in VT aerosols. Of the remaining targeted aerosol constituents reported for 3R4F, comparisons were not made between VT and 3R4F for chromium, nickel, glycidol, lead (VT3-intense), menthol, and ammonia (VT5). Chromium was not detected, nickel was BLOD/BLOQ in both products, ammonia and lead (VT3-intense) were not different from background (NDFB), and glycidol data was not reported since an appropriate test method was not available at the time of this study. As expected, PG and VG were higher in VT vs. 3R4F as they are primary ingredients in JUULpods. As so few aerosol constituents in VT5 and VT3 aerosols were above LOD/LOQ, estimated values (as outlined in the methods) were used to provide a comparison to quantified aerosol constituents in 3R4F smoke. These aerosol constituent values and % Differences are preceded by a "$\leq$" symbol in Table 5 and Supplementary Table S16. Every aerosol constituent in 3R4F mainstream cigarette smoke included here for comparison was quantifiable, except for chromium, nickel, and menthol. Correspondingly, every aerosol constituent in VT that could be compared to reported values for the 3R4F mainstream cigarette smoke (excluding PG, VG, and water) was reduced. Reductions ranged from $\geq$92.22% (furfural; VT3, NI) to 99.99% (propylene oxide; VT5, intense). Notably, quantifiable levels of formaldehyde were reduced from between 96.16% (VT3-NI) and 99.20% (VT5-intense) and acetaldehyde was reduced by $\geq$99% (VT5 and VT3, both regimes). Differences in reductions from 3R4F between the aerosol generated using intense and NI puffing regimens were within 3%, excepting water. Average reductions for aerosol constituents (excluding PG, VG and water) from VT5 and VT3 were $\geq$98.59% (Table 5 and Supplementary Table S16).

Similar to the data for 3R4F mainstream cigarette smoke, not all of the 53 aerosol constituents included in this analysis were reported in the literature for IQOS tobacco flavor heat sticks. Excluding benzoic acid, cotinine (VT5), and nornicotine, none of the chemicals unavailable for comparison were detectable in VT aerosol. Twenty-seven out of thirty-one aerosol constituents reported for the IQOS tobacco heat stick were found at detectable levels. Only 5 of these aerosol constituents were detectable in VT5 and 6 in VT3 (formaldehyde, acetaldehyde [VT3], glycerol, nicotine, propylene glycol, and water). Comparisons were not made for 4-aminobiphenyl (uncertain comparison), ammonia (VT5; NDFB), cadmium (all products BLOD/BLOQ), chromium (all products BLOD/BLOQ), glycidol (not recorded), nickel (all products BLOD/BLOQ), and lead (VT3; NDFB). Similar to the comparison with 3R4F mainstream cigarette smoke, to provide a comparison to aerosol constituents in IQOS, estimated values based on LOD and LOQ were employed for the remainder of JUUL VT5 aerosol constituents, where IQOS aerosol constituent levels were available and quantifiable. All VT5 aerosol constituents which were compared to IQOS were reduced, save glycerol and propylene glycol. Aerosol constituent reductions ranged from $\geq$66.39% (VT5) for 1-aminonaphthalene to $\geq$99.98% (VT5) for diacetyl. Average aerosol constituent reductions (excluding PG, VG, and water) for VT5 and VT3 were $\geq$89.12% (Table 5 and Supplementary Table S16).

3.2. Menthol

The aerosol constituent levels of Me5 and Me3 as compared to 3R4F and IQOS is outlined in Table 6 and Supplementary Table S17. Two of the 53 aerosol constituents (carbon monoxide and gold) were not tested in Menthol 3.0% under either puffing regimen. Me5 contained 10/53 and 8/53 quantifiable aerosol constituents under the intense and NI regimes, respectively, while Me3 contained 8/53 and 7/53. Estimated aerosol constituents are indicated by "$\leq$". Of the aerosol constituents not reported for the mainstream smoke from the 3R4F reference cigarette, only benzoic acid and nornicotine (Me5) were detected.

Table 6. Comparison of Aerosol Constituents in Me5 aerosol Generated Using the Intense Puffing Regime vs. Comparator Products.

Aerosol Constituent	Aerosol Constituents Normalized by Nicotine (mg/mg)					% Difference		
	Me5		3R4F		IQOS	vs. 3R4F		vs. IQOS
	NI	Intense	NI	Intense	Intense	NI	Intense	Intense
1-Aminonaphthalene (CA)	$\leq 2.25 \times 10^{-08}$ [a]	$\leq 2.25 \times 10^{-08}$ [a]	2.02×10^{-05}	1.46×10^{-05}	7.11×10^{-08}	$\downarrow\geq 99.89$	$\downarrow\geq 99.85$	$\downarrow\geq 68.30$
1-Butanol	$\leq 7.45 \times 10^{-04}$ [a]	$\leq 4.33 \times 10^{-04}$ [a]	N/A	N/A	N/A	N/A	N/A	N/A
1,3-Butadiene (CA)	$\leq 7.85 \times 10^{-05}$ [a]	$\leq 7.85 \times 10^{-05}$ [a]	5.40×10^{-02}	5.03×10^{-02}	2.19×10^{-04}	$\downarrow\geq 99.85$	$\downarrow\geq 99.84$	$\downarrow\geq 64.17$
2-Aminonaphthalene (CA)	$\leq 9.80 \times 10^{-09}$ [a]	$\leq 9.80 \times 10^{-09}$ [a]	1.25×10^{-05}	9.10×10^{-06}	$\leq 2.89 \times 10^{-08}$	$\downarrow\geq 99.92$	$\downarrow\geq 99.89$	NC
4-Aminobiphenyl (CA)	$\leq 3.94 \times 10^{-09}$ [a]	$\leq 3.94 \times 10^{-09}$ [a]	2.19×10^{-06}	1.96×10^{-06}	$\leq 4.21 \times 10^{-08}$	$\downarrow\geq 99.82$	$\downarrow\geq 99.80$	NC
Acetaldehyde (CA)	$\leq 5.35 \times 10^{-04}$ [a]	**2.30×10^{-04}**	8.50×10^{-01}	9.24×10^{-01}	1.69×10^{-01}	$\downarrow\geq 99.94$	**$\downarrow 99.98$**	$\downarrow 99.86$
Acetyl Propionyl	$\leq 4.36 \times 10^{-05}$ [a]	$\leq 2.54 \times 10^{-05}$ [a]	4.21×10^{-02}	4.62×10^{-02}	N/A	$\downarrow\geq 99.90$	$\downarrow\geq 99.95$	N/A
Acrolein	$\leq 2.44 \times 10^{-04}$ [a,b]	$\leq 1.29 \times 10^{-04}$ [a,b]	7.58×10^{-02}	9.05×10^{-02}	7.56×10^{-03}	$\downarrow\geq 99.68$	$\downarrow\geq 99.86$	$\downarrow\geq 98.30$
Acrylonitrile (CA)	$\leq 7.95 \times 10^{-05}$ [a]	$\leq 4.63 \times 10^{-05}$ [a]	7.45×10^{-03}	1.03×10^{-02}	1.82×10^{-04}	$\downarrow\geq 98.93$	$\downarrow\geq 99.55$	$\downarrow\geq 74.56$
Ammonia	NDFB	NDFB	1.37×10^{-02}	1.71×10^{-02}	1.14×10^{-02}	NC	NC	NC
Anabasine	$\leq 3.43 \times 10^{-04}$ [a]	$\leq 3.43 \times 10^{-04}$ [a]	N/A	N/A	N/A	N/A	N/A	N/A
Anatabine	$\leq 2.84 \times 10^{-04}$ [a]	$\leq 2.84 \times 10^{-04}$ [a]	N/A	N/A	N/A	N/A	N/A	N/A
Benzene (CA)	$\leq 4.24 \times 10^{-05}$ [a]	$\leq 2.46 \times 10^{-05}$ [a]	4.79×10^{-02}	4.46×10^{-02}	5.29×10^{-04}	$\downarrow\geq 99.91$	$\downarrow\geq 99.94$	$\downarrow\geq 95.34$
Benzo(a)pyrene (CA)	$\leq 1.46 \times 10^{-07}$ [a]	$\leq 1.46 \times 10^{-07}$ [a]	8.97×10^{-06}	7.59×10^{-06}	1.07×10^{-06}	$\downarrow\geq 98.37$	$\downarrow\geq 98.07$	$\downarrow\geq 86.29$
Benzoic Acid	**8.17×10^{-01}**	**7.48×10^{-01}**	N/A	N/A	N/A	N/A	N/A	N/A
Benzyl Acetate	$\leq 7.45 \times 10^{-04}$ [a]	$\leq 4.33 \times 10^{-04}$ [a]	N/A	N/A	N/A	N/A	N/A	N/A
β-Nicotyrine	$\leq 9.24 \times 10^{-05}$ [a,b]	**8.49×10^{-05}**	N/A	N/A	N/A	N/A	N/A	N/A
Cadmium (CA)	$\leq 1.30 \times 10^{-07}$ [a]	$\leq 7.59 \times 10^{-08}$ [a]	3.49×10^{-05}	4.68×10^{-05}	$\leq 2.89 \times 10^{-07}$	$\downarrow\geq 99.63$	$\downarrow\geq 99.84$	NC
Carbon Monoxide	$\leq 1.44 \times 10^{-01}$ [a]	$\leq 1.44 \times 10^{-01}$ [a]	$1.42 \times 10^{+01}$	$1.53 \times 10^{+01}$	4.91×10^{-01}	$\downarrow 98.99$	$\downarrow\geq 99.06$	$\downarrow\geq 70.57$
Chromium	NDFB	NDFB	$\leq 1.25 \times 10^{-06}$	$\leq 6.84 \times 10^{-06}$	$\leq 4.55 \times 10^{-07}$	NC	NC	NC
Copper	$\leq 1.63 \times 10^{-06}$ [a]	$\leq 9.48 \times 10^{-07}$ [a]	N/A	N/A	N/A	N/A	N/A	N/A
Cotinine	$\leq 1.01 \times 10^{-04}$ [b]	$\leq 5.87 \times 10^{-05}$ [b]	N/A	N/A	N/A	N/A	N/A	N/A
Crotonaldehyde (CA)	$\leq 7.73 \times 10^{-05}$ [a]	$\leq 4.50 \times 10^{-05}$ [a]	1.41×10^{-02}	2.99×10^{-02}	2.68×10^{-03}	$\downarrow\geq 99.45$	$\downarrow\geq 99.85$	$\downarrow\geq 98.32$
Diacetyl	$\leq 2.16 \times 10^{-05}$ [a]	$\leq 1.26 \times 10^{-05}$ [a]	1.64×10^{-01}	1.69×10^{-01}	5.42×10^{-02}	$\downarrow\geq 99.99$	$\downarrow\geq 99.99$	$\downarrow\geq 99.98$
Diethylene Glycol	$\leq 2.98 \times 10^{-03}$ [a]	$\leq 1.73 \times 10^{-03}$ [a]	N/A	N/A	N/A	N/A	N/A	N/A
Ethyl Acetate	$\leq 7.45 \times 10^{-04}$ [a]	$\leq 8.32 \times 10^{-04}$ [a,b]	N/A	N/A	N/A	N/A	N/A	N/A
Ethyl Acetoacetate	$\leq 5.96 \times 10^{-05}$ [a]	$\leq 3.47 \times 10^{-05}$ [a]	N/A	N/A	N/A	N/A	N/A	N/A
Ethylene Glycol	$\leq 2.71 \times 10^{-03}$ [b]	$\leq 7.05 \times 10^{-04}$ [b]	N/A	N/A	N/A	N/A	N/A	N/A
Formaldehyde (CA)	**4.41×10^{-04}**	**4.88×10^{-04}**	3.52×10^{-02}	4.38×10^{-02}	3.76×10^{-03}	$\downarrow 98.75$	$\downarrow 98.89$	$\downarrow 87.03$
Furfural	$\leq 1.04 \times 10^{-03}$ [a]	$\leq 6.07 \times 10^{-04}$ [a]	1.84×10^{-02}	1.49×10^{-02}	N/A	$\downarrow\geq 94.34$	$\downarrow\geq 95.92$	N/A
Glycerol	**$1.38 \times 10^{+01}$**	**$1.29 \times 10^{+01}$**	$3.85 \times 10^{+00}$	$1.36 \times 10^{+00}$	$3.26 \times 10^{+00}$	$2.58 \times 10^{+02}$	$8.47 \times 10^{+02}$	$2.94 \times 10^{+02}$
Glycidol (CA)	NR	NR	N/A	8.84×10^{-04}	9.24×10^{-04}	NC	NC	NC
Gold	$\leq 1.74 \times 10^{-07}$ [a]	$\leq 1.74 \times 10^{-07}$ [a]	N/A	N/A	N/A	N/A	N/A	N/A
Isoamyl Acetate	$\leq 1.19 \times 10^{-03}$ [a]	$\leq 6.94 \times 10^{-04}$ [a]	N/A	N/A	N/A	N/A	N/A	N/A
Isobutyl Acetate	$\leq 7.45 \times 10^{-04}$ [a]	$\leq 4.33 \times 10^{-04}$ [a]	N/A	N/A	N/A	N/A	N/A	N/A
Isobutyraldehyde	$\leq 2.05 \times 10^{-05}$ [a]	$\leq 5.17 \times 10^{-05}$ [b]	N/A	N/A	N/A	N/A	N/A	N/A
Isoprene (CA)	$\leq 1.12 \times 10^{-04}$ [a]	$\leq 1.12 \times 10^{-04}$ [a]	4.10×10^{-01}	4.02×10^{-01}	1.74×10^{-03}	$\downarrow\geq 99.97$	$\downarrow\geq 99.97$	$\downarrow\geq 93.59$
Lead	$\leq 5.52 \times 10^{-07}$ [b]	$\leq 6.57 \times 10^{-07}$ [b]	1.31×10^{-05}	1.79×10^{-05}	$\leq 2.77 \times 10^{-06}$	$\downarrow\geq 95.79$	$\downarrow\geq 96.33$	NC
Menthol	**2.02×10^{-01}**	**1.87×10^{-01}**	$\leq 1.32 \times 10^{-05}$	$\leq 4.76 \times 10^{-06}$	$2.17 \times 10^{+00}$	NC	NC	$\downarrow 91.37$
Methyl Acetate	$\leq 8.94 \times 10^{-04}$ [a]	$\leq 5.20 \times 10^{-04}$ [a]	N/A	N/A	N/A	N/A	N/A	N/A
Myosmine	$\leq 5.30 \times 10^{-04}$ [a]	$\leq 5.30 \times 10^{-04}$ [a]	N/A	N/A	N/A	N/A	N/A	N/A
n-Butyraldehyde	$\leq 4.36 \times 10^{-05}$ [a]	$\leq 2.54 \times 10^{-05}$ [a]	4.05×10^{-02}	4.67×10^{-02}	2.21×10^{-02}	$\downarrow\geq 99.89$	$\downarrow\geq 99.95$	$\downarrow\geq 99.89$
Nickel	$\leq 1.13 \times 10^{-05}$ [a,b]	$\leq 6.57 \times 10^{-06}$ [a,b]	$\leq 5.41 \times 10^{-07}$	$\leq 7.41 \times 10^{-06}$	$\leq 4.55 \times 10^{-07}$	NC	NC	NC
Nicotine	**$1.00 \times 10^{+00}$**	**$1.00 \times 10^{+00}$**	$1.00 \times 10^{+00}$	$1.00 \times 10^{+00}$	$1.00 \times 10^{+00}$	N/A	N/A	N/A
Nicotine-N-Oxide	$\leq 2.16 \times 10^{-04}$ [a]	$\leq 1.26 \times 10^{-04}$ [a]	N/A	N/A	N/A	N/A	N/A	N/A
NNK (CA)	$\leq 1.87 \times 10^{-07}$ [a]	$\leq 1.09 \times 10^{-07}$ [a]	1.61×10^{-04}	1.48×10^{-04}	4.88×10^{-06}	$\downarrow\geq 99.88$	$\downarrow\geq 99.93$	$\downarrow\geq 97.77$
NNN (CA)	$\leq 1.22 \times 10^{-07}$ [a]	$\leq 7.11 \times 10^{-08}$ [a]	1.87×10^{-04}	1.69×10^{-04}	1.13×10^{-05}	$\downarrow\geq 99.93$	$\downarrow\geq 99.96$	$\downarrow\geq 99.37$
Nornicotine	**5.48×10^{-04}** [a]	**5.48×10^{-05}** [a]	N/A	N/A	N/A	N/A	N/A	N/A
Propionic Acid	$\leq 1.94 \times 10^{-03}$ [b]	$\leq 1.13 \times 10^{-03}$ [b]	N/A	N/A	N/A	N/A	N/A	N/A
Propylene Glycol	**$5.14 \times 10^{+00}$**	**$5.05 \times 10^{+00}$**	4.21×10^{-03}	1.36×10^{-02}	3.25×10^{-01}	$1.22 \times 10^{+05}$	$3.70 \times 10^{+04}$	$1.45 \times 10^{+03}$
Propylene Oxide (CA)	$\leq 3.87 \times 10^{-05}$ [a]	$\leq 2.26 \times 10^{-05}$ [a]	4.23×10^{-01}	5.03×10^{-01}	1.23×10^{-04}	$\downarrow\geq 99.99$	$\downarrow\geq 99.99$	$\downarrow\geq 81.69$
Toluene	$\leq 1.52 \times 10^{-04}$ [a]	$\leq 8.84 \times 10^{-05}$ [a]	7.32×10^{-02}	7.69×10^{-02}	1.98×10^{-03}	$\downarrow\geq 99.79$	$\downarrow\geq 99.89$	$\downarrow\geq 95.53$
Water	**$2.82 \times 10^{+00}$**	**$2.47 \times 10^{+00}$**	$4.21 \times 10^{+00}$	$8.36 \times 10^{+00}$	$2.45 \times 10^{+01}$	$\downarrow 33.00$	$\downarrow 70.40$	$\downarrow 89.92$
						Average [c] $\downarrow 99.24$	Average [c] $\downarrow 99.40$	Average [c] $\downarrow 89.00$

BLOD = Below limit of quantitation; BLOQ = Below limit of quantitation; NDFB = Not different from background; NC = Not comparable; NR = Not reported; $\downarrow$ = less than. Note: Values preceded with "$\geq$" are based on estimated values. Bolded values were quantifiable. CA = carcinogen. Analytes in italics were tested under a single modified puffing regimen: 70 mL puff, 3 s duration, 30 s interval. [a] analyte was BLOD for all three 50 puff blocks. [b] analyte was BLOQ for all three 50 puff blocks. [a,b] analyte was either BLOD or BLOQ for all three 50 puff blocks. [c] excluding PG, VG, and water.

Comparisons to 3R4F were not made for ammonia (Me5), chromium, glycidol, menthol, and nickel. Chromium was not detected, nickel was BLOD/BLOQ in the aerosols of both products, ammonia was NDFB in Me5 aerosols, and the 3R4F reference cigarette is not a mentholated product. The primary ingredients propylene glycol and glycerol were higher in Me vs. 3R4F mainstream cigarette smoke. Estimated values were again used for aerosol constituents BLOD/BLOQ to provide comparison. Every constituent in Me aerosols which could be compared to 3R4F mainstream cigarette smoke (excluding PG, VG, and water) was reduced. Reductions ranged from ≥88.65% (ammonia; Me3-NI) to 99.99% (propylene oxide; Me5-intense). Quantifiable levels of formaldehyde were reduced by ≥98.75% and acetaldehyde was reduced by ≥99.90%. As with VT aerosol constituents, differences in reductions from 3R4F mainstream cigarette smoke between the intense and NI regimens were within 3%, excepting lead (Me3-based on estimated values) and water. Average reductions (excluding PG, VG, and water) for Me aerosol constituents were ≥97.47% (Table 6 and Supplementary Table S17).

Of the aerosol constituents not reported in IQOS, benzoic acid, β-Nicotyrine (Me5-intense), ethylene glycol (Me5-intense), cotinine (Me3-intense) and nornicotine were detected in Me5. Twenty seven of 31 aerosol constituents reported for mIQOS were present at detectable levels. Seven of these aerosol constituents (acetaldehyde (intense), formaldehyde, glycerol, menthol, nicotine, propylene glycol and water) were detected in JUUL menthol products. Comparisons were not made for 1,3-butadiene (Me3; uncertain comparison), 2-aminonaphthlene (all products BLOD/BLOQ), 4-aminobiphenyl (uncertain comparison), ammonia (Me5; NDFB), cadmium (all products BLOD/BLOQ), chromium (all products BLOD/BLOQ), isoprene (Me3; uncertain comparison), lead (all products BLOD/BLOQ), nickel (all products BLOD/BLOQ) and glycidol (not recorded). All of the JUUL menthol aerosol constituents which were compared to IQOS were reduced save the primary ingredients glycerol and propylene glycol. Reductions in aerosol constituents ranged from ≥61.36% for acrylonitrile (Me3) to ≥99.98% (Me5) for diacetyl. Average aerosol constituent reductions (excluding PG, VG, and water) for Me aerosols were ≥89.00% (Table 6 and Supplementary Table S17).

4. Discussion

4.1. JUUL Aerosol Characterization

The product characterization in this study was focused on 53 aerosol constituents included in draft and final FDA guidance for the tobacco industry [36,37]. Aerosol generation, collection and chemical analysis were performed by ISO 17,025 accredited CROs with validated methodology (Supplemental Table S1). Across all JUUL flavors, nicotine concentrations and puffing regimes, only 10 of the 53 aerosol constituents were measured above their limit of quantification in at least one flavor and one puffing regime. These aerosol constituents included: acetaldehyde, benzoic acid, β-nicotyrine, cotinine, formaldehyde, glycerol, nicotine, nornicotine, propylene glycol, and water. As expected, the primary e-liquid ingredients in JUUL products (i.e., nicotine, propylene glycol, glycerol, and benzoic acid) were detected in the aerosol.

Of the 10 quantifiable aerosol constituents generated from the JUUL system, acetaldehyde and formaldehyde were the only two quantifiable carbonyls. While formaldehyde was present at detectable levels in the aerosols of all four JUUL aerosols, acetaldehyde was only quantifiable in VT3, Me5 (intense), and Me3 (intense). Acetaldehyde ranged from ≤0.022 to 0.044 μg/puff, and formaldehyde ranged from 0.022 to 0.087 μg/puff. The concentrations of these carbonyl compounds in e-cigarette aerosols have been documented previously in multiple publications. Their presence is hypothesized to result mainly from the thermal degradation of the primary e-liquid ingredients PG and VG, the mechanism of which was summarized in Flora et al. [54]. This reaction is reported to correlate with coil temperature [24,55]. Conversely, the JUUL device has a regulated temperature [34]. This is likely a main contributor to the low levels of carbonyl compounds observed in JUUL aerosols. The measured values for formaldehyde and acetaldehyde, in all samples, did not significantly increase in the end puff block over the middle puff blocks. All analytes that were quantifiable in the end puff blocks were also quantifiable in the previous puff blocks.

The following nicotine-related impurities/degradants were above the limit of quantification for one or more of the JUUL system products: nornicotine (ranged from

0.017–0.024 μg/puff), cotinine (0.00094 to 0.0066 μg/puff), and β-nicotyrine (Me5-intensive, 0.013 μg/puff). The JUUL system e-liquids are formulated with USP-grade nicotine. The USP nicotine standard has acceptance criteria of not more than 0.3% for each nicotine-related compound and 0.8% for total impurities for nicotine-related compounds (i.e., anabasine, anatabine, nicotyrine, cotinine, myosmine, nicotine-N-oxide, and nornicotine) [56], so low levels of nicotine-related impurities in aerosols are expected. For all JUUL system aerosols tested, the summation of nicotine-related compounds did not exceed 0.25% of the measured nicotine concentration, which was well below the USP purity standard.

4.2. Aerosol Constituent Comparison between JUUL System and 3R4F Reference Cigarette

Across all flavors and nicotine concentrations, aerosol constituents were reduced in JUUL products relative to 3R4F mainstream cigarette smoke. Of those which could be compared, all aerosol constituents in the JUUL system (except PG, VG, and water) were present at substantially lower levels relative to the levels in 3R4F mainstream cigarette smoke. Average aerosol constituent reductions (excluding nicotine, PG, VG, and water) for all four JUULPods tested, regardless of puffing regimes, were greater than 98% lower than levels in 3R4F mainstream tobacco smoke (Table 4).

Acrolein, crotonaldehyde, diacetyl, and n-Butyraldehyde were quantifiable in 3R4F mainstream cigarette smoke and were BLOD in JUUL product aerosols. Carbonyl aerosol constituent reductions for JUUL aerosols were > 80% compared to levels in 3R4F mainstream cigarette smoke. The aromatic amines 1-amnonaphthalene, 2-aminonaphthalene, and 4-aminobiphenyl were not detected in JUUL system aerosols. Using aerosol constituent values estimated based on method LOD, these constituents are >99% lower than those in 3R4F. The volatile organic compounds 1,3-butadiene, acrylonitrile, benzene, isoprene, and toluene while quantifiable in 3R4F mainstream cigarette smoke were BLOD in JUUL aerosols. Estimated aerosol constituent values indicate a >99% reduction in the JUUL system relative to 3R4F mainstream cigarette smoke. None of the six metals tested (cadmium, chromium, copper, gold, nickel, and lead) were above LOQ in JUUL system aerosols. Cadmium and gold were BLOD, chromium was NDFB; and copper, nickel, and lead were alternately BLOD or BLOQ across flavors, nicotine concentrations, and puff blocks. Estimated aerosol constituent values indicate a >86% reduction in comparison to levels in 3R4F mainstream smoke.

The primary e-liquid ingredients PG and VG were found to be higher in the JUUL systems versus the 3R4F reference cigarette. PG and VG are common base ingredients in ENDS products [22] and are generally used as humectants in combusted cigarettes [57]. Although no comparison was made for benzoic acid, it is assumed to be higher in JUUL aerosols.

4.3. Aerosol Constituent Comparison between JUUL System and IQOS

All aerosol constituents compared in the JUUL System, excepting PG and VG, were present at lower levels relative to the yields in IQOS aerosol, resulting in > 88% average reductions (excluding PG, VG, and water) across the JUUL products (Table 4). Across all flavors and nicotine concentrations, the carbonyls (i.e., acetaldehyde, acrolein, crotonaldehyde, diacetyl, formaldehyde, and n-Butyraldehyde) were substantially lower for JUUL aerosols than for IQOS. They were all quantifiable in IQOS aerosols, while only acetaldehyde and formaldehyde were above LOQ in JUUL system aerosols.

1-Aminonaphthalene (quantifiable in IQOS aerosols) was not detectable in JUUL system aerosols. 2-Aminonaphthalene and 4-Aminobiphenyl were quantifiable in IQOS tobacco flavor aerosols and were BLOD in all JUUL system aerosols. Volatile organic compounds ([VOCs]; 1,3-butadiene, acrylonitrile, benzene, isoprene, and toluene) were quantifiable in IQOS aerosols but were BLOD in JUUL aerosols. Using estimated values indicates a >61% reduction of these VOCs in JUUL system aerosols compared to IQOS. Among the six metals targeted for analysis in JUUL system aerosols, none were quantifiable. In the available literature on IQOS aerosols, there was information on four metals. Of these, only lead was quantifiable in IQOS tobacco flavor aerosol. Based on estimated values, lead was reduced by ≥86% to ≥97% in JUUL aerosols vs. IQOS.

4.4. Aerosol Constituent Comparison with Literature Values

Reilly et al. reported aerosol yields of formaldehyde, acetaldehyde, and nicotine from a 5% nicotine, tobacco flavored JUULPod and device [32]. In that study, aerosols were generated from 10 puffs using a 75 mL puff, a duration of 2.5 s, and an interpuff interval of 30 s. Four replicates were collected. This regime is similar to the NI puffing regime used in the current study. The authors reported quantifiable results for formaldehyde and nicotine while acetaldehyde was reported as BLOD. Normalized by nicotine yield, Reilly et al. reported 1.47×10^{-03} (mg/mg) of formaldehyde. Comparatively, formaldehyde is reported as 9.82×10^{-04} (mg/mg) for VT5, under NI puffing, in the current study. Collaborative study results from the CORESTA e-vapour subgroup have shown intra-laboratory differences from ~40% to 150% for the determination of formaldehyde in the aerosol produced using the same device. The difference between the reported values between these two studies falls within this range.

Talih et al. reported aerosol yields for a range of compounds from an unnamed US market 5% nicotine JUULPod and device [31]. Aerosol samples were collected from 15 puffs using a 66.7 mL puff, a duration of 4 s and an interpuff interval of 10 s. Although puff volume and duration are comparable to the NI regime of our study, Talih and colleagues employed a much shorter interpuff interval. We therefore make a comparison to our intense regime. They reported quantifiable results for nicotine and carbonyl compounds including formaldehyde and acetaldehyde. Normalized by nicotine yield, the authors reported 3.13×10^{-03} mg/mg and 2.98×10^{-03} of mg/m) for formaldehyde and acetalde-hyde respectively. Comparatively, our values were 3.50×10^{-04} mg/mg (formaldehyde) and $\leq 2.16 \times 10^{-04}$ mg/mg (acetaldehyde) for VT5, under intense puffing. The values reported by Talih et al. are approximately 10 times higher than the current study, likely due to differences in the puffing regimen. Nevertheless, the values reported by Talih et al. represent a 92.9% and 99.9% reduction in acetaldehyde and formaldehyde, respectively, compared to the 3R4F intense data shown in Table 5. These results indicate that even when testing with a 10 s interpuff interval, JUULSystem aerosol is significantly reduced in acetaldehyde and formaldehyde compared to the mainstream smoke of a 3R4F cigarette.

4.5. Study Limitations

One of the study limitations was use of aerosol constituent levels reported in the literature. There were instances where the aerosol constituent level reported in the literature for the mainstream smoke from the 3R4F reference cigarette or IQOS were quantifiable at a level that was below the analytical method LOD or LOQ used for JUUL Product aerosol constituent measurements. In these cases, no direct comparison of aerosol constituent levels could be made. This was the case for three JUUL aerosol analytes: 4-aminobiphenyl (VT5), 1,3-Butadiene (Me3), and isoprene (Me3). Another limitation was that although originally planned, a validated fit for purpose method for the determination of glycidol was not available when this study was initiated, therefore aerosol constituent levels of glycidol are not reported. Another limitation is that targeted analysis is not comprehensive, but only quantifies pre-determined aerosol constituents. There may be additional constituents present in JUUL system aerosols that were not in the targeted list. To address this, we performed a non-targeted analysis that is described in a companion paper to this work. The values presented in this study were generated using standardized machine smoking and puffing regimes, which are appropriate for comparisons between products. The results are not intended to be representative of possible human exposure.

5. Conclusions

In this study, we measured 53 tobacco-related HPHCs and chemicals in aerosols from four JUUL products currently available on the US market and compared them to cigarette smoke and the aerosol from a heated tobacco product. Of the 53 toxicants, only 10 were quantifiable in at least one JUUL product aerosol and puffing regime. Average reductions (excluding the primary e-liquid ingredients PG, VG, and water) for all JUUL flavors tested were reduced by more than 98% compared to the 3R4F, and 88% compared to IQOS.

The data indicated that although JUUL aerosols have detectable levels of known degradants of PG/VG (acetaldehyde and formaldehyde) and nicotine-related compounds, the vast majority of tobacco-related HPHCs were not detectable in JUUL aerosols. The low levels of HPHCs in the JUUL system aerosol are likely due to the tightly controlled temperature regulation of the JUUL system designed to reduce byproducts of combustion.

Supplementary Materials: The following are available online at https://www.mdpi.com/article/10.3390/separations8100168/s1, Supplementary Tables S1–S17. Table S1: Supplementary Methods. Table S2: Method LOD and LOQ for Me3 analysis. Table S3: Method LOD and LOQ for VT5, VT3 and Me5 analysis. Table S4: Comparison of HPHC and Chemical Intense Aerosol Levels in Virginia Tobacco 5.0% and Smoke Levels in 3R4F Reference Cigarette. Table S5: Comparison of HPHC and Chemical Non-Intense Aerosol Levels in Virginia Tobacco 5.0% and Smoke Levels in 3R4F Reference Cigarette. Table S6: Comparison of HPHC and Chemical Non-Intense Aerosol Levels in Virginia Tobacco 5.0% and Smoke Levels in IQOS regular. Table S7: Comparison of HPHC and Chemical Intense Aerosol Levels in Virginia Tobacco 3.0% and Smoke Levels in 3R4F Reference Cigarette. Table S8: Comparison of HPHC and Chemical Non-Intense Aerosol Levels in Virginia Tobacco 3.0% and Smoke Levels in 3R4F Reference Cigarette. Table S9: Comparison of HPHC and Chemical Intense Aerosol Levels in Virginia Tobacco 3.0% and Aerosol Levels in IQOS Regular. Table S10: Comparison of HPHC and Chemical Intense Aerosol Levels in 10enthol 5.0% and Smoke Levels 3R4F Reference Cigarette. Table S11: Comparison of HPHC and Chemical Non-Intense Aerosol Levels in Menthol 5.0% and Smoke Levels in 3R4F Reference Cigarette. Table S12: Comparison of HPHC and Chemical Intense Aerosol Levels in Menthol 5.0% and Aerosol in IQOS Menthol. Table S13: Comparison of HPHC and Chemical Intense Aerosol Levels in Menthol 13.0% and Smoke Levels 3R4F Reference Cigarette. Table S14: Comparison of HPHC and Chemical Non-Intense Aerosol Levels in Menthol 3.0% and Smoke Levels in 3R4F Reference Cigarette. Table S15: Comparison of HPHC and Chemical Intense Aerosol Levels in Menthol 3.0% and Aerosol Levels in IQOS Menthol. Table S16: Comparison of HPHCs in VT3 aerosol vs. comparator products. Table S17: Comparison of HPHCs in Me3 aerosol vs. comparator products.

Author Contributions: Conceptualization, I.G.G. and X.C.; formal analysis, C.Y., B.H. and X.C.; data curation, X.C.; writing—original draft preparation, P.C.B. and X.C.; writing—review and editing, I.G.G. and M.J.O.; visualization—P.C.B.; supervision, I.G.G.; project administration, X.C. All authors have read and agreed to the published version of the manuscript.

Funding: This research received no external funding.

Institutional Review Board Statement: Not applicable.

Informed Consent Statement: Not applicable.

Data Availability Statement: Data is contained within the article and Supplementary Materials.

Acknowledgments: The authors acknowledge the technical assistance of David Cook and Adam Ozvald with the manuscript preparation.

Conflicts of Interest: P.C.B., X.C., I.G.G., R.W., C.Y., B.H. and M.J.O. are employees of JUUL Labs, Inc.

References

1. United States Food & Drug Administration. *Harmful and Potentially Harmful Constituents in Tobacco Products and Tobacco Smoke, Established List 2012*; FDA: Montgomery, MD, USA, 2021.

2. United States Food & Drug Administration A. Reporting Harmful and Potentially Harmful Constituents in Tobacco Products and Tobacco Smoke Under Section 904(a) (3) of the Federal Food, Drug, and Cosmetic Act: Draft Guidance for Industry. Available online: https://www.fda.gov/regulatory-information/search-fda-guidance-documents/reporting-harmful-and-potentially-harmful-constituents-tobacco-products-and-tobacco-smoke-under (accessed on 27 September 2021).

3. Warren, G.W.; Alberg, A.J.; Kraft, A.S.; Cummings, K.M. The 2014 Surgeon General's report: "The Health consequences of smoking-50 years of progress": A paradigm shift in cancer care. *Cancer* **2014**, *120*, 1914–1916. [CrossRef]

4. Farsalinos, K.E.; Polosa, R. Safety evaluation and risk assessment of electronic cigarettes as tobacco cigarette substitutes: A systematic review. *Ther. Adv. Drug Saf.* **2014**, *5*, 67–86. [CrossRef] [PubMed]

5. El-Hellani, A.; Salman, R.; El-Hage, R.; Talih, S.; Malek, N.; Baalbaki, R.; Karaoghlanian, N.; Nakkash, R.; Shihadeh, A.; Saliba, N.A. Nicotine and carbonyl emissions from popular electronic cigarette products: Correlation to liquid composition and design characteristics. *Nicotine Tob. Res.* **2016**, *20*, 215–223. [CrossRef]

6. Laugesen, M. Nicotine and toxicant yield ratings of electronic cigarette brands in New Zealand. *N. Z. Med. J.* **2015**, *128*, 77–82.

7. Tayyarah, R.; Long, G.A. Comparison of select analytes in aerosol from e-cigarettes with smoke from conventional cigarettes and with ambient air. *Regul. Toxicol. Pharmacol.* **2014**, *70*, 704–710. [CrossRef] [PubMed]

8. Goniewicz, M.L.; Knysak, J.; Gawron, M.; Kosmider, L.; Sobczak, A.; Kurek, J.; Prokopowicz, A.; Jabłońska-Czapla, M.; Rosik-Dulewska, C.; Havel, C.; et al. Levels of selected carcinogens and toxicants in vapour from electronic cigarettes. *Tob. Control* **2013**, *23*, 133–139. [CrossRef]

9. Flora, J.W.; Wilkinson, C.T.; Wilkinson, J.W.; Lipowicz, P.J.; Skapars, J.A.; Anderson, A.; Miller, J.H. Method for the determination of carbonyl compounds in e-cigarette aerosols. *J. Chromatogr. Sci.* **2017**, *55*, 142–148. [CrossRef]

10. Farsalinos, K.E.; Kistler, K.A.; Pennington, A.; Spyrou, A.; Kouretas, D.; Gillman, G. Aldehyde levels in e-cigarette aerosol: Findings from a replication study and from use of a new-generation device. *Food Chem. Toxicol.* **2018**, *111*, 64–70. [CrossRef]

11. Gillman, I.G.; Pennington, A.S.; Humphries, K.E.; Oldham, M.J. Determining the impact of flavored e-liquids on aldehyde production during Vaping. *Regul. Toxicol. Pharmacol.* **2020**, *112*, 104588. [CrossRef] [PubMed]

12. Margham, J.; McAdam, K.; Forster, M.; Liu, C.; Wright, C.; Mariner, D.; Proctor, C. Chemical composition of aerosol from an e-cigarette: A quantitative comparison with cigarette smoke. *Chem. Res. Toxicol.* **2016**, *29*, 1662–1678. [CrossRef]

13. Kim, H.-J.; Shin, H.-S. Determination of tobacco-specific nitrosamines in replacement liquids of electronic cigarettes by liquid chromatography—Tandem mass spectrometry. *J. Chromatogr. A* **2013**, *1291*, 48–55. [CrossRef] [PubMed]

14. Long, G.A. Comparison of select analytes in exhaled aerosol from e-cigarettes with exhaled smoke from a conventional cigarette and exhaled breaths. *Int. J. Environ. Res. Public Health* **2014**, *11*, 11177–11191. [CrossRef]

15. Farsalinos, K.E.; Gillman, I.G.; Melvin, M.S.; Paolantonio, A.R.; Gardow, W.J.; Humphries, K.E.; Brown, S.E.; Poulas, K.; Voudris, V. Nicotine levels and presence of selected tobacco-derived toxins in tobacco flavoured electronic cigarette refill liquids. *Int. J. Environ. Res. Public Health* **2015**, *12*, 3439–3452. [CrossRef]

16. Pellegrino, R.M.; Tinghino, B.; Mangiaracina, G.; Marani, A.; Vitali, M.; Protano, C.; Osborn, J.F.; Cattaruzza, M.S. Electronic cigarettes: An evaluation of exposure to chemicals and fine particulate matter (PM). *Ann. Ig.* **2012**, *24*, 279–288.

17. Beauval, N.; Antherieu, S.; Soyez, M.; Gengler, N.; Grova, N.; Howsam, M.; Hardy, E.M.; Fischer, M.; Appenzeller, B.M.; Goossens, J.-F.; et al. Chemical Evaluation of electronic cigarettes: Multicomponent analysis of liquid refills and their corresponding aerosols. *J. Anal. Toxicol.* **2017**, *41*, 670–678. [CrossRef]

18. Marco, E.; Grimalt, J.O. A rapid method for the chromatographic analysis of volatile organic compounds in exhaled breath of tobacco cigarette and electronic cigarette smokers. *J. Chromatogr. A* **2015**, *1410*, 51–59. [CrossRef]

19. Williams, M.; Villarreal, A.; Bozhilov, K.; Lin, S.; Talbot, P. Metal and silicate particles including nanoparticles are present in electronic cigarette cartomizer fluid and aerosol. *PLoS ONE* **2013**, *8*, e57987. [CrossRef]

20. Lerner, C.A.; Sundar, I.K.; Watson, R.; Elder, A.; Jones, R.; Done, D.; Kurtzman, R.; Ossip, D.J.; Robinson, R.; McIntosh, S.; et al. Environmental health hazards of e-cigarettes and their components: Oxidants and copper in e-cigarette aerosols. *Environ. Pollut.* **2015**, *198*, 100–107. [CrossRef]

21. National Academies of Sciences, Engineering, and Medicine. *Public Health Consequences of E-Cigarettes*; National Academies Press: Washington, DC, USA, 2018.

22. Royal College of Physicians. *Nicotine Without Smoke: Tobacco Harm Reduction*; RCP: London, UK, 2016.

23. Gillman, I.; Kistler, K.; Stewart, E.; Paolantonio, A. Effect of variable power levels on the yield of total aerosol mass and formation of aldehydes in e-cigarette aerosols. *Regul. Toxicol. Pharmacol.* **2016**, *75*, 58–65. [CrossRef] [PubMed]

24. Uchiyama, S.; Noguchi, M.; Sato, A.; Ishitsuka, M.; Inaba, Y.; Kunugita, N. Determination of thermal decomposition products generated from e-cigarettes. *Chem. Res. Toxicol.* **2020**, *33*, 576–583. [CrossRef] [PubMed]

25. Geiss, O.; Bianchi, I.; Barrero-Moreno, J. Correlation of volatile carbonyl yields emitted by e-cigarettes with the temperature of the heating coil and the perceived sensorial quality of the generated vapours. *Int. J. Hyg. Environ. Health* **2016**, *219*, 268–277. [CrossRef]

26. Kosmider, L.; Sobczak, A.; Fik, M.; Knysak, J.; Zaciera, M.; Kurek, J.; Goniewicz, M. Carbonyl compounds in electronic cigarette vapors: Effects of nicotine solvent and battery output voltage. *Nicotine Tob. Res.* **2014**, *16*, 1319–1326. [CrossRef] [PubMed]

27. Jensen, R.P.; Luo, W.; Pankow, J.F.; Strongin, R.M.; Peyton, D.H. Hidden formaldehyde in e-cigarette aerosols. *N. Engl. J. Med.* **2015**, *372*, 392–394. [CrossRef]

28. Zelinkova, Z.; Wenzl, T. Influence of battery power setting on carbonyl emissions from electronic cigarettes. *Tob. Induc. Dis.* **2020**, *18*, 77. [CrossRef] [PubMed]

29. Ward, A.M.; Yaman, R.; Ebbert, J.O. Electronic nicotine delivery system design and aerosol toxicants: A systematic review. *PLoS ONE* **2020**, *15*, e0234189. [CrossRef]

30. Son, Y.; Bhattarai, C.; Samburova, V.; Khlystov, A. Carbonyls and carbon monoxide emissions from electronic cigarettes affected by device type and use patterns. *Int. J. Environ. Res. Public Health* **2020**, *17*, 2767. [CrossRef]

31. Talih, S.; Salman, R.; El-Hage, R.; Karam, E.; Karaoghlanian, N.; El-Hellani, A.; Saliba, N.; Shihadeh, A. Characteristics and toxicant emissions of JUUL electronic cigarettes. *Tob. Control* **2019**, *28*, 678–680. [CrossRef] [PubMed]
32. Reilly, S.M.; Bitzer, Z.T.; Goel, R.; Trushin, N.; Richie, J.P. Free radical, carbonyl, and nicotine levels produced by juul electronic cigarettes. *Nicotine Tob. Res.* **2019**, *21*, 1274–1278. [CrossRef]
33. Crosswhite, M.R.; Bailey, P.C.; Jeong, L.N.; Lioubomirov, A.; Yang, C.; Ozvald, A.; Jameson, J.B.; Gillman, I.G. Non-targeted chemical characterization of Juul Virginia tobacco flavored aerosols using liquid and gas chromatography. *Separations* **2021**, *8*, 130. [CrossRef]
34. Alston, B. Measurement of temperature regulation performance of the JUUL nicotine salt pod system. In Proceedings of the CORESTA, Hamburg, Germany, 6–10 October 2019.
35. Talih, S.; Salman, R.; Soule, E.; El-Hage, R.; Karam, E.; Karaoghlanian, N.; El-Hellani, A.; Saliba, N.; Shihadeh, A. Electrical features, liquid composition and toxicant emissions from 'pod-mod'-like disposable electronic cigarettes. *Tobacco Control* **2021**. [CrossRef]
36. United States Food & Drug Administration. *Premarket Tobacco Product Applications for Electronic Nictotine Delivery Systems: Guidance for Industry-Draft Guidance*; FDA: Montgomery, MD, USA, 2016.
37. United States Food & Drug Administration. *Premarket Tobacco Product Applications for Electronic Nicotine Delivery Systems: Guidance for Industry*; FDA: Montgomery, MD, USA, 2019.
38. United States Food & Drug Administration. *Harmful and Potentially Harmful Constituents in Tobacco Products, Established List, Proposed Additions, Request for Comments*; FDA: White Oak, MD, USA, 2019.
39. International Council on Harmonisation. *Harmonised Tripartite Guideline: Validation of Analytical Procedures, Text and Methodology*; ICH: Geneva, Switzerland, 2005.
40. ISO 8454:2007. *Cigarettes-Determination of Carbon Monoxide in the Vapour Phase of Cigarette Smoke-NDIR Method*; ISO: Geneva, Switzerland, 2007.
41. ISO 20768:2018. *Vapour Products-Routine Analytical Vaping Machine-Definitions and Standard Conditions*; ISO: Geneva, Switzerland, 2018.
42. Guthery, W. Emissions of toxic carbonyls in an electronic cigarette. *Beiträge Tabakforschung International* **2016**, *27*, 30–37. [CrossRef]
43. Tayyarah, R. *2014 Electronic Cigarette Aerosol Parameters Study*; Coresta, March 2015; Available online: https://www.coresta.org/sites/default/files/technical_documents/main/ECIG-CTR_ECigAerosolParameters-2014Study_March2015.pdf (accessed on 28 September 2021).
44. ISO 3308:2012. *Routine Analytical Cigarette-Smoking Machine-Definitions and Standard Conditions*; ISO: Geneva, Switzerland, 2012.
45. ISO 20778:2018. Cigarettes. In *Routine Analytical Cigarette Smoking Machine: Definitions and Standard Conditions with an Intense Smoking Regime*; ISO: Geneva, Switzerland.
46. Jaccard, G.; Djoko, D.T.; Korneliou, A.; Stabbert, R.; Belushkin, M.; Esposito, M. Mainstream smoke constituents and in vitro toxicity comparative analysis of 3R4F and 1R6F reference cigarettes. *Toxicol. Rep.* **2019**, *6*, 222–231. [CrossRef]
47. Moldoveanu, S.; Hudson, A.; Harrison, A. The Determination of Diacetyl and Acetylpropionyl in Aerosols From Electronic Smoking Devices Using Gas Chromatography Triple Quad Mass Spectrometry. *Contrib. Tob. Res.* **2017**, *27*, 145–153. [CrossRef]
48. Pappas, R.S.; Fresquez, M.R.; Martone, N.; Watson, C.H. Toxic metal concentrations in mainstream smoke from cigarettes available in the USA. *J. Anal. Toxicol.* **2014**, *38*, 204–211. [CrossRef] [PubMed]
49. Schaller, J.-P.; Keller, D.; Poget, L.; Pratte, P.; Kaelin, E.; McHugh, D.; Cudazzo, G.; Smart, D.; Tricker, A.R.; Gautier, L.; et al. Evaluation of the tobacco heating system Part 2: Chemical composition, genotoxicity, cytotoxicity, and physical properties of the aerosol. *Regul. Toxicol. Pharmacol.* **2016**, *81*, S27–S47. [CrossRef] [PubMed]
50. U.S. Food and Drug Administration. Technical Project Review of IQOS PMTA Submission. Available online: https://www.fda.gov/media/144701/download (accessed on 17 February 2020).
51. Cox, S.; Goniewicz, M.L.; Kosmider, L.; McRobbie, H.; Kimber, C.; Dawkins, L. The Time course of compensatory puffing with an electronic cigarette: Secondary analysis of real-world puffing data with high and low nicotine concentration under fixed and adjustable power settings. *Nicotine Tob. Res.* **2021**, *23*, 1153–1159. [CrossRef] [PubMed]
52. Dawkins, L.; Cox, S.; Goniewicz, M.; McRobbie, H.; Kimber, C.; Doig, M.; Kośmider, L. 'Real-world' compensatory behaviour with low nicotine concentration e-liquid: Subjective effects and nicotine, acrolein and formaldehyde exposure. *Addiction* **2018**, *113*, 1874–1882. [CrossRef] [PubMed]
53. Ashton, H.; Stepney, R.; Thompson, J.W. Self-titration by cigarette smokers. *BMJ* **1979**, *2*, 357–360. [CrossRef] [PubMed]
54. Flora, J.W.; Meruva, N.; Huang, C.B.; Wilkinson, C.T.; Ballentine, R.; Smith, D.C.; Werley, M.S.; McKinney, W.J. Characterization of potential impurities and degradation products in electronic cigarette formulations and aerosols. *Regul. Toxicol. Pharmacol.* **2016**, *74*, 1–11. [CrossRef] [PubMed]
55. Wang, P.; Chen, W.; Liao, J.; Matsuo, T.; Ito, K.; Fowles, J.; Shusterman, D.; Mendell, M.; Kumagai, K. A device-independent evaluation of carbonyl emissions from heated electronic cigarette solvents. *PLoS ONE* **2017**, *12*, e0169811. [CrossRef]
56. United States Pharmacopeia. *Nicotine USP 43 NF 38-3152*; United States Pharmacopeial Convention, Inc.: Rockville, MD, USA, 2020.
57. World Health Organization. *Standard Operating Procedure for Determination of Humectants in Cigarette Tobacco Filler*; WHO: Geneva, Switzerland, 2015.

 *separations*

Article

Non-Targeted Chemical Characterization of JUUL Virginia Tobacco Flavored Aerosols Using Liquid and Gas Chromatography

Mark R. Crosswhite *, Patrick C. Bailey, Lena N. Jeong, Anastasia Lioubomirov, Clarissa Yang, Adam Ozvald, J. Brian Jameson and I. Gene Gillman

Juul Labs, Inc., Washington, DC 20004, USA; patrick.bailey@juul.com (P.C.B.); lena.jeong@juul.com (L.N.J.); anastasia.lioubomirov@juul.com (A.L.); clarissa.yang@juul.com (C.Y.); adam.ozvald@juul.com (A.O.); j.brian@juul.com (J.B.J.); gene.gillman@juul.com (I.G.G.)
* Correspondence: mark.crosswhite@juul.com

Citation: Crosswhite, M.R.; Bailey, P.C.; Jeong, L.N.; Lioubomirov, A.; Yang, C.; Ozvald, A.; Jameson, J.B.; Gillman, I.G. Non-Targeted Chemical Characterization of JUUL Virginia Tobacco Flavored Aerosols Using Liquid and Gas Chromatography. *Separations* **2021**, *8*, 130. https://doi.org/10.3390/separations8090130

Academic Editors: Fadi Aldeek and Alena Kubatova

Received: 13 July 2021
Accepted: 17 August 2021
Published: 24 August 2021

Abstract: The chemical constituents of JUUL Virginia Tobacco pods with 3.0% and 5.0% nicotine by weight (VT3 and VT5) were characterized by non-targeted analyses, an approach to detect chemicals that are not otherwise measured with dedicated methods or that are not known beforehand. Aerosols were generated using intense and non-intense puffing regimens and analyzed by gas chromatography electron ionization mass spectrometry and liquid chromatography electrospray ionization high resolving power mass spectrometry. All compounds above 0.7 µg/g for GC–MS analysis or above 0.5 µg/g for LC–HRMS analysis and differing from blank measurements were identified and semi-quantified. All identifications were evaluated and categorized into five groups: flavorants, harmful and potentially harmful constituents, extractables and/or leachables, reaction products, and compounds that could not be identified/rationalized. For VT3, 79 compounds were identified using an intense puffing regimen and 69 using a non-intense puffing regimen. There were 60 compounds common between both regimens. For VT5, 85 compounds were identified with an intense puffing regimen and 73 with a non-intense puffing regimen; 67 compounds were in common. For all nicotine concentrations, formulations and puffing regimens, reaction products accounted for the greatest number of compounds (ranging from 70% to 75%; 0.08% to 0.1% by mass), and flavorants comprised the second largest number of compounds (ranging from for 15% to 16%; 0.1 to 0.2% by mass). A global comparison of the compounds detected in JUUL aerosol to those catalogued in cigarette smoke indicated an approximate 50-fold decrease in chemical complexity. Both VT3 and VT5 aerosols contained 59 unique compounds not identified in cigarette smoke.

Keywords: JUUL; aerosol; non-targeted analysis; chemical characterization; ENDS; e-cigarette; GC–MS; LC–HRMS

1. Introduction

In light of the fact that combustible cigarette (CC) smoking is the number one cause of preventable death in the world [1], causing over 8 million deaths each year and comprising 30% of cancer-related deaths overall [2], numerous public health agencies have developed programs with the intent of both preventing smoking initiation and promoting smoking cessation [3,4]. Tobacco product regulations, such as the U.S. Food and Drug Administration (FDA) "Family Smoking Prevention and Tobacco Control Act", were enacted to help protect public health by regulating the manufacturing, distribution, and marketing of tobacco products [5]. There are, however, many smokers who are not likely to quit in the near term [6]. With this in mind, in 2017, the FDA announced the Comprehensive Plan for Tobacco and Nicotine Regulation ("Comprehensive Plan"), which recognizes that nicotine is delivered on a continuum of risk and seeks to render cigarettes and other combustible tobacco products minimally or non-addictive through the creation of a very low nicotine

91

cigarette product standard [7]. This continuum of risk places combustible cigarettes at the highest risk and nicotine replacement therapies at the lowest risk of the harm spectrum. This continuum of risk is based in large part on evidence that nicotine, while addictive, is not itself responsible for serious disease and death in cigarette smokers [8]. Rather, it is the combination of thousands of other chemical constituents present in the smoke of CCs [9].

The chemical composition of CC smoke has been well studied [10] and much is known about the harmfulness of smoking [1]. This is due partly to the length of time researchers have been studying smoking and its negative effects on the population [1]. A great deal of our current understanding of the harmfulness of smoking cigarettes is founded on an understanding of the compounds which are produced during the tobacco combustion process. Temperatures at the center of a burning cigarette range from 600 to 900 °C. In addition to combustion byproducts, the high temperatures involved in the production of mainstream smoke result in incomplete combustion, which causes thermal degradation of the tobacco plant materials, paper, and non-tobacco ingredients. Cigarette smoke is a highly complex mixture of >5000 constituents including carcinogenic, mutagenetic and respiratory toxicants [10,11]. Therefore, numerous public health organizations have developed lists of toxicants in the smoke of CCs [12–14]. The FDA's established list in particular contains 93 harmful or potentially harmful constituents (HPHCs) recognized in tobacco products [14].

Byproducts of tobacco combustion are responsible for many of the toxicants present in cigarette smoke [11]. Electronic Nicotine Delivery System (ENDS) products are designed to operate below combustion temperatures, which may reduce the toxicant production when compared with CCs [9]. ENDS represent a fundamentally different approach to delivering nicotine versus CC [15] and may reduce the harm potential of nicotine-containing aerosols [16,17]. The e-liquid of ENDS products is compositionally different than a tobacco cigarette. Plant material and paper are not present nor combusted in an ENDS product. Therefore, many HPHCs present in mainstream smoke are either not present in ENDS product aerosols, or present at significantly lower levels than CC smoke, i.e., ammonia, aromatic amines, carbon monoxide, tobacco-specific nitrosamines, polycyclic aromatic hydrocarbons, and volatile organic compounds [15,16,18–21]. However, questions regarding the complexity of ENDS aerosols and the possible existence of unique constituents which pose potential harm apart from the known HPHCs of CCs remain [22,23].

Bentley and colleagues performed non-targeted analysis (NTA) of the IQOS heated tobacco product which showed that IQOS aerosol is much less complex than CC smoke [24]. The JUUL System, like IQOS, operates below combustion temperatures and, unlike IQOS, has no tobacco paper or plant material (excluding tobacco-derived nicotine and tobacco-derived flavorants). Therefore, we hypothesized that the chemical composition of JUUL System aerosols would also be less complex and contain fewer HPHCs than CC smoke.

Regulators and other public health organizations are also interested in addressing the concern that there may be uniquely harmful constituents in the aerosol of ENDS products. On 5 August 2019, the FDA proposed the addition of 19 ENDS-specific chemicals to the HPHC list [14]. These proposed additions include some flavorants, glycerol (vegetable glycerin or VG), and propylene glycol (PG), which are constituents for many ENDS products. While the expanded HPHC list included some ENDS-specific constituents, it does not fully encompass the range of ingredients used in ENDS formulations or constituents present in ENDS aerosols. Formulation ingredients are unique to each ENDS product and multiple ingredients may react to form a wide range of reaction products, similar to reaction chemistry previously observed with a wide range of food grade flavors [25]. Erythropel et al. reported data that indicated that primary constituents and flavorant aldehydes may react with each other to form acetals of PG and VG [26–28]. It is plausible that in addition to these reactions, which require mixing and time, other reactions could take place as a result of heating, contact with the heating coil material, or increased exposure to water and oxygen during aerosol formation. In addition, flavorants and extractable and leachable (E&L) compounds that could be transferred to the aerosol should be evaluated as well.

The majority of analytical work on ENDS has focused on targeting known chemicals of interest (i.e., the analytes being determined are known beforehand and the methods are tailored to detect those chemicals) based on changes to the device, formulation, power, temperature or sampling approaches [29]. For example, Vreeke and colleagues used targeted GC–MS and nuclear magnetic resonance to measure the amount of dihydroxyacetone generated from ENDS as a function of operating wattages [30]. Electron spin resonance has been used to determine free radicals in ENDS aerosols with GC–MS analysis employed to characterize flavor components [31]. Kosmider et al. used GC–MS to determine nicotine emission from ENDS as a function of PG and VG composition as well as device power [32], while Zhao and colleagues used liquid chromatography coupled to tandem mass spectrometry and electron spin resonance (ESR) targeted methods to detect stable/short lived reactive oxygen species (ROS) and found a substantial influence of e-cig brand, e-liquid flavor, puffing pattern and operational voltage on ROS levels [33]. Farsalinos and Gillman published a review focused on carbonyl emission from ENDS devices which covered 32 English-language studies in which 22 puffing variations and 9 trapping approaches are represented [34]. While many known chemicals have been targeted for analysis, these methods may yet leave unsampled portions, leading to potential gaps in the understanding of ENDS aerosol composition. Apart from the work of a few research groups [35–37], little has been published on the NTA of nicotine-containing aerosol from combustible and heated tobacco products, and even less on ENDS aerosols [38–41].

To address the potential gaps in understanding left by targeted analysis, NTA characterization requires the ability to capture, detect, and identify compounds relevant to a specific chemical space. The non-targeted analyses performed in this study were designed to span physicochemical properties from non-volatile to volatile and non-polar to polar chemicals [24]. This is best achieved using at least a set of two complimentary non-targeted methods. GC–MS methods are well suited and widely accepted for analysis of volatile sensory, flavor, and aroma compounds [42] as demonstrated by its application to coffee [43], fermentation products [44], bread [45], Scotch Whisky [46], wine [47], tobacco [48], olive oil [49] and electronic cigarettes [50]. GC–MS methods are well suited for non-targeted analyses because amount estimations without a reference standard are possible and because there are standardized electron ionization spectral libraries that facilitate compound identification [51]. For example, Krüsemann, using heat-assisted diffusion of volatiles in tobacco products to the gas phase, measured results against large databases in order to create a library of flavor compounds [48]. GC–MS alone is not sufficient for characterizing all portions of ENDS aerosol because there are liquid droplets also present in the aerosol and they may contain chemicals that are not amendable to GC–MS analysis. This is due to the fact that as the bulk liquids are heated, mixed with air and converted into an aerosol, there is a potential for non-volatile and higher-molecular-weight compounds to be present in the microdroplets of the carrier [36]. Therefore, for a thorough characterization, there is a need for an LC–MS-based non-targeted approach as well.

This combination of GC–MS and LC–MS-based non-targeted methodologies was employed in the current study to detect chemicals not included in the FDA's list of 93 HPHCs, the proposed additions to the HPHC list, or elsewhere in the literature. The approach was designed with the intent to detect a large portion of the composition of JUUL aerosol and to elucidate constituents in common with, and unique from, the smoke of CCs using published data. The thorough chemical characterization of JUUL System aerosols described here is part of a multi-path approach to generate the necessary data for a toxicological risk assessment of JUUL use.

2. Materials and Methods

Two semi-quantitative non-targeted analyses were implemented to compliment targeted methods in order to provide a more complete list of aerosol constituents. The NTA methods were developed to be suitable for the detection and identification of chemicals from a broad chemical space (Figure 1). It is necessary to consider the properties of the anticipated compounds (i.e., volatile, non-volatile, polar, non-polar, etc.) in order to achieve a robust and thorough characterization, and to minimize sample preparation/manipulation. Therefore, the samples were collected and analyzed without any matrix removal steps, which minimized analyte loss and enabled the capture of a full range of diverse chemical constituents. This was achieved by collecting aerosols through a quartz filter pad and chilled impinger containing ethanol. After collection, the pad and ethanol were combined to extract the pad contents, and the resulting solution was sampled without any further manipulation or dilution.

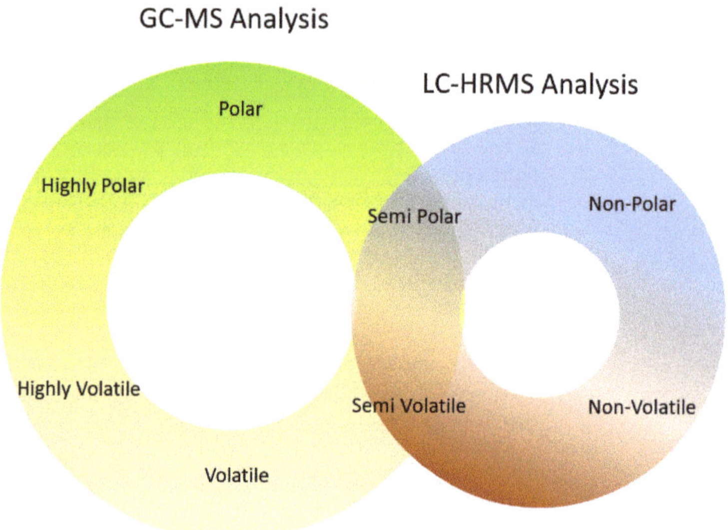

Figure 1. Technical Coverage of Constituent Properties in JUUL Aerosol Chemical Space.

Two complimentary analytical techniques were employed: gas chromatography electron ionization mass spectrometry and liquid chromatography electrospray ionization high resolving power mass spectrometry. These techniques were optimized to complement each other to provide maximal coverage of the chemicals potentially present in the aerosol of ENDS products. The GC–MS method was optimized for the detection of volatile and polar compounds [39]. Concurrently, the LC–HRMS method was optimized for detection of non/semi-volatile compounds and non/semi-polar compounds [52] (Figure 1). These techniques employ comparison to a known amount of an internal standard to achieve a quantity estimation across multiple compounds. It should be noted, however, that while these complementary methods cover a broad chemical space, not all chemicals present in the aerosol are detectable under these methods. For example, chemicals such as metals, non-ionizable compounds, compounds that are not amenable to chromatography, or compounds outside of the mass-to-charge (M/Z) scan range of the mass detector cannot be detected.

Software platforms were utilized for data processing of both GC–MS and LC–HRMS data to facilitate compound detection and identification. Agilent MassHunter Unknowns Analysis software (GC–MS) (Santa Clara, CA, USA) and Thermo Compound Discoverer version 3.0 (LC–HRMS) (Waltham, MA, USA) were employed to search both commercial and custom mass spectral libraries to identify potential aerosol constituents. This workflow was employed to investigate the chemical composition of JUUL aerosols with the aim of providing semi-quantitative information of constituents and evaluating the relative complexity of JUUL aerosols compared to cigarette smoke [10].

In this study, differential analyses based on nine ($n = 9$) collection replicates of each of the nicotine strengths (3.0% and 5.0%) and each collection condition (intense and non-intense) were used to characterize compounds differing from collection blanks. This method relies on the application of statistical tools to extract the relevant information from a large and highly complex dataset [53,54]. Due to the large number of variables in non-targeted analyses relative to the number of samples, these tools are imperative to avoid misinterpretation of instrument and collection artifacts as sample relevant compounds [55].

To allow for a more complete understanding of aerosol chemistry, tentatively identified ENDS aerosol analytes were rationalized into defined groups (Table 1). Compounds listed in the Flavor Extract Manufacturers Association (FEMA) flavor ingredient library [56] were labeled as flavorants. Compounds which are listed by the FDA as HPHCs in Tobacco Products and Tobacco Smoke: Established List [14] were labeled as HPHCs. Any compounds which are commonly found to migrate from packaging materials of consumer products [57–59] were labeled as extractables and leachables (E&L). Compounds proposed to be a result of chemical reactions, except when the product is an HPHC, were labeled as reaction products (Figure 2). All other compounds which were not able to be identified or rationalized were assigned to group 5.

Table 1. Classification of Rational Compound Origin Identification.

Group Number	Group Definition
1	Flavorants
2	HPHCs listed in Tobacco Products and Tobacco Smoke: Established List
3	Extractables and leachables
4	Any compound resulting from a chemical reaction
5	Compounds unable to be identified/rationalized

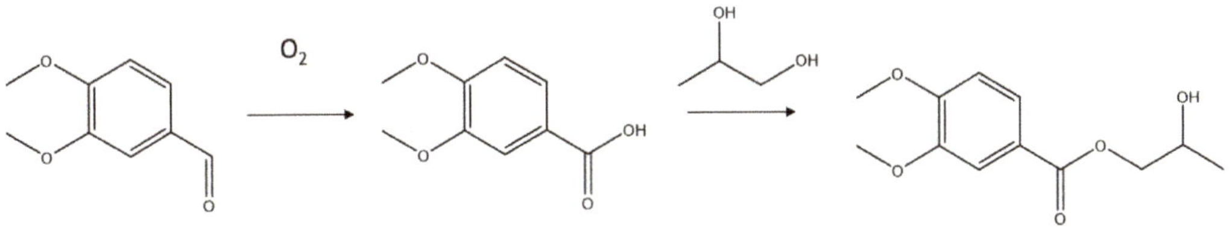

Figure 2. Examples of Possible Chemical Pathways Used for Rationalizing a Reaction Product.

All samples were equilibrated to room temperature if removed from environmental chambers. Prior to aerosol collection, the JUULpod was attached to a fully charged JUUL device. All aerosol collections were performed on Borgwaldt LX20 (Hamburg, Germany Part# 12000820) linear puffing machines and the JUUL System was inserted into a custom pad holder containing a 55 mm glass fiber filter pad (GFFP) (Part # 9703-9024, Whatman) to trap non-volatile compounds during aerosol collection. A chilled impinger ($-5 \pm 5\,°C$) containing 10 µg/mL 6-methylcoumarin (Sigma-Aldrich P/N W269905-100G-K), in 200 proof ethanol (Pharmco-Aaper P/N 111000200) as a trapping solvent was used in conjunction with the GFFP. The device was oriented at a 45° angle, with the battery end downward. The GFFP was extracted in the impinger solution and shaken for 30 to 60 min. The resulting solution was subjected to GC–MS and LC–HRMS analysis. Additionally, aerosol blanks were collected using an open port on the puffing machine and by puffing room air across the filter pad and through the impinger concurrently with sample collections. The pad and trapping solution for the blank were treated and analyzed the same as samples but were differentiated from samples during data processing. Three production batches and three replicates from each batch were analyzed for a total of nine replicates per JUUL product.

As recommended in the FDA's guidance on premarket tobacco product applications for ENDS, non-intense and intense puffing regimens were used for all aerosol measurements performed with the JUUL System [60]. The non-intense puffing regimen followed the ISO 20768:2018 standard (square wave 55 mL over 3 s every 30 s). The intense puffing regimen used a square wave puff volume of 110 mL over 6 s every 30 s (Table 2). As opposed to discrete puffing blocks, an end of life (EOL) study was carried out to determine the number of machine puffs needed to fully deplete a JUULpod. Sample collections were set to achieve 85–90% of total EOL aerosol yield [61]. End of life testing was performed for both non-intense and intense puffing regimens, using the same product batches included in this study to ensure that the product performance was consistent with the samples being tested. The puffing was done in sequential 50-puff blocks, with the devices removed from the smoking machine every 50 puffs and inverted three times to settle the e-liquid on the wick. Devices were replaced every 50 puffs with fully charged devices. The device mass loss was determined by weighing the device prior to and after each 50-puff block. The EOL for a given pod was defined as the 50-puff collection where device mass loss was <10 mg/50 puffs for each replicate. The results of the EOL study showed that the intense regimen required about 2/3 the number of puffs to deplete a JUULpod compared to non-intense puffing, despite having 2-fold the puff volume. The six second duration for intense puffs was chosen because the JUUL device heating cuts off after six seconds of continuous puffing.

Table 2. Aerosol Generation for Samples.

Group	Puff Volume (mL)	Duration (s)	Puff Interval (s)	Puff Count (n)
VT3				
intense	110	6	30	225
non-intense	55	3	30	315
VT5				
intense	110	6	30	225
non-intense	55	3	30	320

LC–HRMS analysis was conducted at and by Juul Labs, Inc. using a Thermo Liquid Chromatograph coupled with Q-Exactive Orbitrap mass spectrometer (Thermo Fisher Scientific, Waltham, MA, USA) in both the full scan and data-dependent acquisition modes. Compounds were separated on a Waters (Milford, MA, USA) BEH C_{18} column (2.1 × 100 mm, 1.7 μm) Part # 186002352 over 26 min. The mobile phase gradient started at 95% 5 mM ammonium acetate in water, reaching 80% methanol at 22 min, followed by a 4 min re-equilibration at starting conditions. Compounds were detected in the positive electrospray ionization (+ESI) mode. Mass spectrometric data were acquired in the full scan mode at 140,000 resolving power from m/z 60 to 800. A pooled mixture of all samples was prepared as a quality control sample to monitor and compensate for time-dependent batch effects as well as for collected data-dependent MS/MS spectra.

Compound Discoverer version 3.0 software (Thermo Fisher Scientific, Waltham, MA, USA) was used to detect compounds and search +ESI spectra compound libraries (Juul Labs, Inc., Washington, DC, USA) custom compound mass spectral library and Thermo mzCloud mass spectral database) to identify the compounds detected by the NTA. All reported constituent amounts were estimated by comparison with the internal standard. All compounds with estimated amounts at or above 0.5 μg/g and a probability value (*p*-value) less than 0.05 were considered sample relevant. A *p*-value less than 0.05 is considered statistically significant (less than a 5% probability that the results are random) [62]. A match factor criterion guideline for the identification confidence levels of compounds identified by LC–HRMS was formulated as shown in Table 3. The final identification confidence levels were assigned after a visual inspection of the mass spectrum.

Table 3. Confidence Levels of Compounds Identified by LC–HRMS NTA.

Confirmed	Comparison to standard reference material
High	High mass accuracy measurement generated a molecular formula which was consistent with a rationalized identification
Medium	Automated search of a library with MS/MS spectral peaks in common
Low	Manual search of online library for best spectral match
NA	Not applicable

GC–MS analyses were conducted at and by Altria Client Services, LLC. (Richmond VA, USA) using a validated method that was included under the site's ISO 17025 accreditation. The scope of this method is to identify new compounds or compounds that increase in concentration over time during stability studies for ENDS products. This provides semi-quantitative results based on the response factor of an analog internal standard. Analyses were performed on an Agilent (Santa Clara, CA, USA) gas chromatograph with electron ionization single quadrupole mass spectrometer (7890 GC/5977 MSD). Compounds were separated using a Restek Stabilwax capillary column (30 m, 0.25 mm ID) over a 27 min temperature gradient starting at 60 °C and held for 1.25 min, followed by 15 °C/min to 210 °C and held for 2 min, finishing by ramping 30 °C/min to 260 °C held for 9 min and mass spectrometric data were acquired in the full scan mode from m/z 60 to 400.

MassHunter Unknowns Analysis software (Agilent, Santa Clara, CA, USA) was used to deconvolute GC–MS full scan EI mass spectra to detect compounds and assign tentative identifications. Compound identifications for peaks in study samples were completed by comparing the mass spectra from the samples to the National Institute of Standards and Technology 2017 mass spectral database and an in-house-developed custom mass spectral library. The custom library consists of mass spectra from matrix matched reference standards, tentative identifications, and previously observed unknown compounds relevant to the sample. All reported compounds from the GC–MS results were detected in the aerosol at greater than 3-fold the estimated amounts in blanks (pad blank, aerosol collection blank, and reagent blank). All reported constituent amounts were estimated by comparison with the internal standard. A match factor criterion guideline for the identification confidence levels of compounds identified by GC–MS was formulated as shown in Table 4. The final identification confidence levels were assigned after a visual inspection of the mass spectrum.

Table 4. Confidence Levels of Compounds Identified by GC–MS NTA.

Confirmed	Identification confirmed by comparison to standard reference material by high and unit mass resolution mass spectrometry
High	A mass spectrum match factor score of 850 to 1000
Medium	A mass spectrum match factor score of 700 to 849
Low	A mass spectrum match factor score of 500 to 699
NA	Not Applicable

Full instrument conditions are outlined in the Extended Methods section of the Supplemental Information.

3. Results

All reported compounds from LC–HRMS and GC–MS NTA results were detected in the aerosol at or above 0.5 µg/g with a *p*-value less than 0.05 [62] for LC–HRMS analysis and at or above 0.7 µg/g and greater than 3-fold the signal in the sample vs. the blank for GC–MS analysis. Nicotine, PG, VG, and benzoic acid are not reported as their concentration exceeded the dynamic range and linear response of the detectors. This rendered it challenging to estimate their amounts while also estimating the amounts of low concentration analytes. Glycidol is also excluded from the NTA results because it is known to form from thermal degradation of glycerol under GC inlet temperatures of 260 °C [63]. The confidence levels of tentative identifications were determined based on the criteria presented in Tables 3 and 4. The identity of each compound was rationalized and categorized into one of the five groups: (1) flavorant, (2) HPHC [14], (3) extractable or leachable, (4) reaction product, or (5) not rationalized (Table 1).

3.1. Virginia Tobacco 3.0% Nicotine (VT3)

A total of 79 compounds were detected in VT3 aerosol collected under the intense puffing regimen, and 69 compounds were detected under the non-intense regimen. The total aerosol constituents tentatively identified in VT3 aerosol (intense and non-intense) using LC–HRMS and GC–MS are outlined in Supplementary Tables S1–S4.

A comparison of GC–MS and LC–HRMS results indicate that the techniques used in this study are generally complementary, with a small overlap of eight compounds detected in both analyses. For both intense and non-intense aerosol samples, these compounds included hydroxyacetone, 2-hydroxypropyl but-3-enoate isomer 1 and isomer 2, veratryl aldehyde, veratryl aldehyde PG acetal isomer 1 and isomer 2, triethyl citrate, and damascenone. The GC–MS method is less susceptible to analyte-specific ionization efficiency when compared to the LC–HRMS method. Therefore, in instances when both LC–HRMS and GC–MS detected the same compound, the GC–MS estimated amounts were used for the calculation in Tables 5 and 6.

Table 5. Summary of Non-Targeted Analyses of VT3 Aerosol Using the Intense Puffing Regimen.

Group Number	Group Name	Average Mass (µg/g)	Average % Aerosol Mass	Number of Compounds	% Number of Compounds
1	Flavorants	1315.0	1315×10^{-4}	13	16
2	HPHCs	ND	ND	ND	ND
3	Extractables and Leachables	30.7	30.7×10^{-4}	5	6
4	Reaction Products	970.9	970.9×10^{-4}	55	70
5	Not Rationalized	16.6	16.6×10^{-4}	6	8
	Total	2333.2	2333.2×10^{-4}	79	100

ND = not detected.

Table 6. Summary of Non-Targeted Analyses of VT3 Aerosol Using the Non-Intense Puffing Regimen.

Group Number	Group Name	Average Mass (µg/g)	Average % Aerosol Mass	Number of Compounds	% Number of Compounds
1	Flavorant	2014.3	2014.3×10^{-4}	11	16
2	HPHCs	ND	ND	ND	ND
3	Extractables and Leachables	46.0	46.0×10^{-4}	5	7
4	Reaction Products	958.8	958.8×10^{-4}	52	75
5	Not Rationalized	0.8	0.8×10^{-4}	1	1
	Total	3019.9	3019.9×10^{-4}	69	100

ND = not detected.

Figure 3 provides a visual summary of the total percent (%) mass breakdown of constituents in the VT3 intense aerosol detected by NTA. In total, the five groups represent 0.23% of the total aerosol mass. The remaining aerosol mass detected by NTA consisted of the major ingredient components PG, VG, nicotine and benzoic acid.

Table 5 summarizes the total number and percent aerosol mass represented by each group of compounds identified in the VT3 intense puffing regimen data. Among the five groups, reaction products (Figure 3) accounted for 70% of the total number of compounds identified. These reaction products however, make up only one-tenth of one percent (0.097%) of the total aerosol mass. Flavorants accounted for 16% of the total number of compounds and 0.13% of the aerosol. Extractables and leachables accounted for 6% of the total number of compounds and a small percentage of the total aerosol mass (0.003%). There were no HPHCs detected for VT3 intense aerosol using these NTA methods—as is the case for all SKUs included in this study; this is due to several reasons. First, some HPHCs are byproducts of complete and incomplete combustion of tobacco plant materials and paper. Given that there is no tobacco plant material—excluding tobacco-derived nicotine and tobacco derived flavorants—nor paper in JUUL products, many HPHCs are not observed at all. Second, some highly volatile or reactive HPHCs, such as formaldehyde, are typically collected with specialized derivatization protocols unlike those used in NTA. Third, some HPHCs that would be detectable by the reported NTA method, for example tobacco-specific nitrosamines NNN and NNK, are found by targeted methods to be absent from or present in concentrations below the limit of detection for these NTA methods [64]. Only a very small percentage of the total aerosol mass (0.0001%) was classified as unknowns or could not be rationalized into a specific group based on tentative identifications.

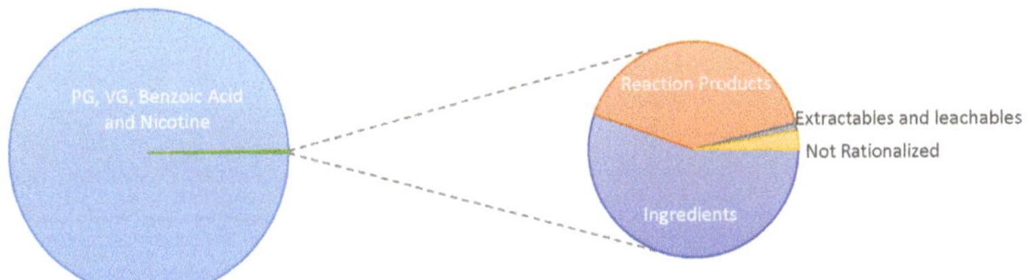

Figure 3. % Total Mass of Aerosol Constituents Detected by NTA Grouped by Type for Aerosol of VT3 Under the Intense Puffing Regimen.

Table 6 summarizes the total number and percent aerosol mass represented by each group of the compounds identified in the VT3 non-intense puffing regimen data. Among the 5 groups, reaction products accounted for 75% of the total number of compounds identified. These reaction products, however, make up only one-tenth of one percent (0.096%) of the total aerosol mass. Flavorants comprised 16% of the total number of compounds and 0.20% of the aerosol mass. Extractables and leachables accounted for 7% of the total number of compounds and a small percentage of the total aerosol mass (0.0046%). HPHCs were not detected (ND) for VT3 non-intense aerosol using these NTA methods and only a very small percentage of the total aerosol mass (0.0001%) were unknowns or could not be rationalized into a specific group based on the tentative identification.

As summarized in Table 7, there were 60 compounds in common between the two puffing topographies. Intense puffing resulted in 19 unique identifications (24% of 79 compounds) while non-intense puffing resulted in 9 unique IDs (13% of 69 compounds). Therefore, the total number of compounds in the aerosol of VT3 was determined to be 88 (60 common identifications + 28 unique identifications). Overall, 42% of compound IDs were confirmed by standard reference material. The remaining 58% compounds were identified with "high" (24%), "medium" (8%), "low" (1%), and "NA" (25%) confidence levels. Among 22 compounds with "NA" confidence level, 13 compounds were classified as nicotine-related degradants. In summary, excluding nicotine, benzoic acid, PG, and VG, the compounds discovered by semi-quantitative NTA methods accounted for approximately 0.2% and 0.3% of the total detected aerosol mass in the intense and non-intense aerosol, respectively.

Although the total number of compounds detected using non-intense topography was smaller than the total number of compounds detected using the intense regimen, the total calculated average mass of collected non-primary-ingredient compounds under non-intense puffing was larger than that for the intense regimen. A difference in the masses collected in the flavorant group accounted for the majority of this inconsistency, and coelution of triethyl citrate with benzoic acid in some analyses was determined to be the primary cause of the inconsistency between intense and non-intense aerosol samples for this group. Prior to the analysis non-intense samples, chromatographic separation between triethyl citrate and benzoic acid was improved, resulting in higher levels of measured triethyl citrate. Overall, approximately 50% of the NTA detected mass was attributed to flavorants. There were six compounds in total that were determined to be associated with E&L, under either puffing regimen. The majority of compounds rationalized as reaction products were associated with PG, VG or nicotine related degradation. A very small percentage of the total aerosol mass (0.0017% for intense and 0.0001% for non-intense puffing topographies) could not be rationalized into a specific group based on the tentative chemical identification.

Table 7. Summary of Non-Targeted Analyses of VT3 Aerosol Using Both Puffing Regimens.

Group Number	Group Name	Intense Number of Compounds	Non-Intense Number of Compounds	Number of Common Compounds	Average Mass (µg/g) Intense	Average Mass (µg/g) Non-Intense
1	Flavorant	13	11	11	1315.0	2014.3
2	HPHCs	ND	ND	ND	ND	ND
3	Extractables and Leachables	5	5	4	30.7	46.0
4	Reaction Products	55	52	45	970.9	958.8
5	Not Rationalized	6	1	0	16.6	0.8
	Total	79	69	60	2333.2	3019.9

ND = not detected.

A global compilation of 5162 compounds in CC smoke catalogued by Rodgman and Perfetti [10] was compared to the 88 compounds detected in VT3 aerosol (Figure 4). Conservatively, this comparison was performed using CAS number, meaning that unless a compound from the NTA was fully identified, it was labeled to be exclusive to JUUL aerosol. Of the 88 compounds detected in VT3, 29 were found to be in common with cigarette smoke and 59 were found to be unique to VT3 (supplemental Table S9). Of the 59 compounds, 44 were termed unique due to lack of CAS number and 24 were classified as nicotine degradants. Table 8 summarizes the total number and aerosol mass represented by each group of the 59 unique compounds in VT3 aerosol. The largest contributing group by mass was flavorant due to the presence of triethyl citrate. Triethyl citrate has been detected in tobacco [65] and in an isolated pyrolysis study, but not detected in tobacco smoke [10,66]. The largest number of compounds exclusive to VT3 aerosols fall into the category of reaction products. Of the 48 reaction products, 19 were nicotine degradants (13 without known structures), 9 were PG/VG degradants and 20 were the product of chemical reactions. The remaining 6 compounds which were not rationalized comprised a very small amount of the total mass (0.0016%). A complete list of all compounds either common with or unique from CC smoke used to generate Figure 4 is presented in Supplementary Table S9.

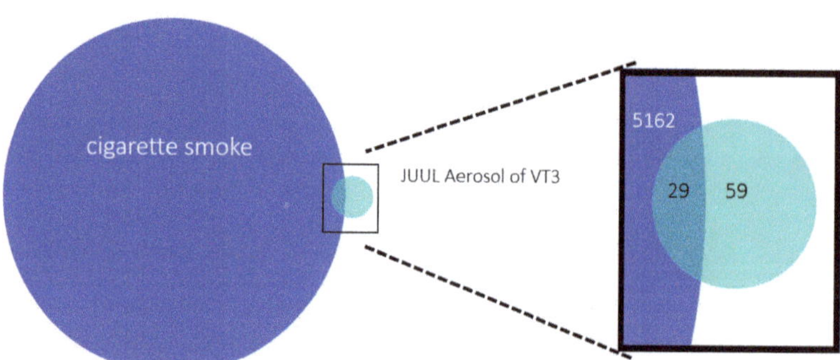

Figure 4. Comparison of the Compounds Detected in VT3 Aerosol to Compounds Found in Smoke of Combustible Cigarettes (Circles are to Scale).

Table 8. Compounds Exclusive to JUUL VT3 Aerosol Compared to Cigarette Smoke.

Group Name	Number of Compounds Unique to VT3 Aerosol	Average Mass (µg/g)	Average % Aerosol Mass
Flavorant	1	1439.4	1439.4×10^{-4}
HPHCs	ND	ND	ND
Extractables and Leachables	4	32.9	32.9×10^{-4}
Reaction Products	48	988.6	988.6×10^{-4}
Not Rationalized	6	16.0	16.0×10^{-4}
Total	59	2476.9	2476.9×10^{-4}

ND = not detected.

3.2. Virginia Tobacco 5.0% Nicotine (VT5)

A total of 85 compounds were detected in VT5 aerosol collected under the intense puffing regimen and 73 compounds were detected under the non-intense puffing regimen. The total aerosol constituents discovered in VT5 aerosol (intense and non-intense) using LC–HRMS and GC–MS are outlined in Supplementary Tables S5–S8.

A comparison of LC–HRMS and GC–MS results indicates that the techniques used in this study are generally complementary, with only ten compounds detected in both analyses. These compounds included hydroxyacetone, 2-hydroxypropyl but-3-enoate isomers 1 and 2, veratryl aldehyde, 1-(1-methyl-5-(pyridine-3-yl)-1H-pyrrol-2-yl) propan-2-one isomer 1 and 2, veratryl aldehyde PG acetal isomers 1 and 2, triethyl citrate, and damascenone for both intense and non-intense aerosol samples. As with the VT3 data, when both LC–HRMS and GC–MS NTA detected the same compounds, the GC–MS estimated amounts were used for calculations in Tables 9 and 10.

Table 9. Summary of Non-Targeted Analyses in VT5 Aerosol Using the Intense Puffing Regimen.

Group Number	Group Name	Average Mass (µg/g)	Average % Aerosol Mass	Number of Compounds	% Number of Compounds
1	Flavorants	975.9	975.9×10^{-4}	13	15
2	HPHCs	ND	ND	ND	ND
3	Extractables and Leachables	41.8	41.8×10^{-4}	3	4
4	Reaction Products	891.6	891.6×10^{-4}	61	72
5	Not Rationalized	38.0	38.0×10^{-4}	8	9
	Total	1947.2	1947.2×10^{-4}	85	100

ND = not detected.

Table 10. Summary of Non-Targeted Analyses in VT5 Aerosol Using the Non-Intense Puffing Regimen.

Group Number	Group Name	Average Mass (µg/g)	Average % Aerosol Mass	Number of Compounds	% Number of Compounds
1	Flavorants	983.7	983.7×10^{-4}	12	16
2	HPHCs	ND	ND	ND	ND
3	Extractables and Leachables	38.1	38.1×10^{-4}	3	4
4	Reaction Products	801.2	801.2×10^{-4}	52	71
5	Not Rationalized	22.0	22.0×10^{-4}	6	8
	Total	1844.9	1844.9×10^{-4}	73	100

ND = not detected.

Figure 5 provides a visual summary of the total percent (%) mass breakdown of constituents in the VT5 intense aerosol detected by NTA. In total, the 5 groups represent less than approximately 0.2% of the total aerosol mass and the rest of the aerosol mass percent detected by NTA is comprised of the major ingredient components PG, VG, nicotine, and benzoic acid.

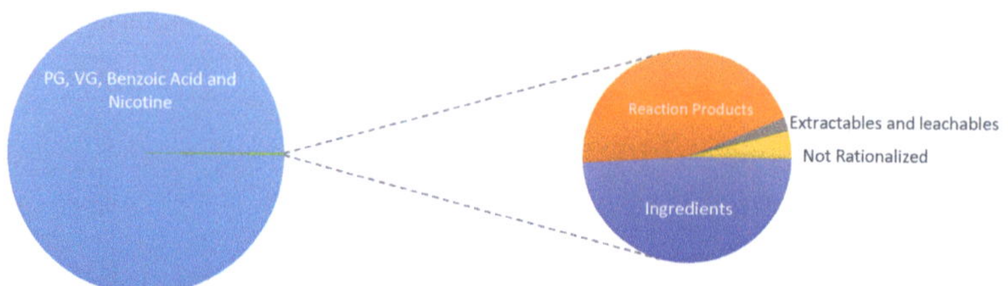

Figure 5. % Total Mass of Constituents by Type in Aerosol of VT5 Under the Intense Puffing Regimen.

Table 9 summarizes the total number and percent aerosol mass represented by each group of compounds identified in the VT5 intense puffing regimen data. Among the five groups, reaction products accounted for 72% of the total number of compounds identified. These reaction products, however, make up less than one-tenth of one percent (0.089%) of the total detected aerosol mass. Flavorants accounted for 15% of the total number of compounds and 0.098% of the detected aerosol mass. E&L accounted for 4% of the total number of compounds and 0.0042% of the total aerosol mass. There were no HPHCs detected for VT5 intense aerosol using these NTA methods and only 0.0038% of the total aerosol mass were unknowns or could not be rationalized into a specific group based on the tentative identification.

Table 10 summarizes the total number and percent aerosol mass represented by each group of the compounds identified in the VT5 non-intense puffing regimen data. Among the five groups, reaction products accounted for 71% of the total number of compounds identified, but less than one-tenth of one percent (0.080%) of the total aerosol mass. Flavorants comprised 16% of the total number of compounds and 0.098% of the detected aerosol mass. Extractables and leachables accounted for 4% of the total number of compounds 0.0038% of the total detected aerosol mass. There was only 0.0022% of the total aerosol mass were unknowns or could not be rationalized into a specific group based on the tentative identification.

As summarized in Table 11, there were 67 compounds in common between the two puffing regimens. Intense puffing resulted in 18 unique identifications (21% of 85 compounds) and non-intense puffing resulted in 6 unique compounds (8% of the 73 compounds). Therefore, the total number of unique compounds in VT5 aerosol was determined to be 91 (67 common identifications + 24 unique IDs). Overall, 47% of compound identifications were confirmed by standard reference material. The remaining 53% of compounds were identified with "high" (20%), "medium" (6%), "low" (2%), and "NA" (25%) confidence levels. Among 23 compounds with "NA" confidence level, 14 compounds were classified as nicotine-related degradants.

Table 11. Summary of Non-Targeted Analyses in Aerosol for Virginia Tobacco 5.0% Using Two Puffing Regimens.

Group Number	Group Name	Intense Number of Compounds	Non-Intense Number of Compounds	Number of Common Compounds	Average Mass (µg/g) Intense	Average Mass (µg/g) Non-Intense
1	Flavorants	13	12	12	975.9	983.7
2	HPHCs	ND	ND	ND	ND	ND
3	Extractables and Leachables	3	3	2	41.8	38.1
4	Reaction Products	61	52	49	891.6	801.2
5	Not Rationalized	8	6	4	38.0	22.0
Total		85	73	67	1947.2	1844.9

ND = not detected.

In summary, the semi-quantitative NTA methods accounted for approximately 0.2% of the total aerosol mass in both intense and non-intense aerosols. The total number of compounds detected using non-intense topography was smaller than the total number of compounds detected using the intense regimen. Most of the total detected mass of both puffing regimens was comprised of flavorants and reaction products. There were four compounds in total that were determined to be associated with E&L, under either puffing regimen. Most reaction products were associated with either PG, VG, or nicotine-related degradation. The small number of remaining compounds were related to E&L or could not be rationalized into one of these categories. Only a very small percentage of the total detected aerosol mass (0.0038% for intense and 0.0022% for non-intense puffing regimens) could not be rationalized into a specific group based on the tentative chemical identification.

As with VT3, the 91 compounds detected in the aerosol of VT5 were compared to the 5162 compounds in CC smoke catalogued by Rodgman and Perfetti (Figure 6). This comparison was also performed using only compounds fully identified with a CAS number, meaning that compounds which were not identified fully were labeled to be exclusive to JUUL aerosol. Of the 91 compounds detected in VT5, 32 were found to be in common with cigarette smoke and 59 were labeled as unique to VT5. Table 12 summarizes the total number and aerosol mass represented by each group of the 59 unique compounds in VT5 aerosol (Table 10). A high percentage of the mass of analytes exclusive to JUUL was related to one flavorant compound, triethyl citrate. According to Rodgman and Perfetti, triethyl citrate has been detected in tobacco and an isolated pyrolysis study, but not detected in tobacco smoke [10]. Reaction products account for the largest number and mass of compounds exclusive to VT5 aerosol. Of the 47 reaction products, 20 were nicotine degradants (14 without known structures), 9 were PG/VG degradants and 18 were the product of other chemical reactions. The remaining 9 compounds which were not rationalized comprised only 0.0039% of the total mass. A complete list of all compounds either common with or unique from CC smoke used to generate Figure 6 is presented in Supplementary Table S10.

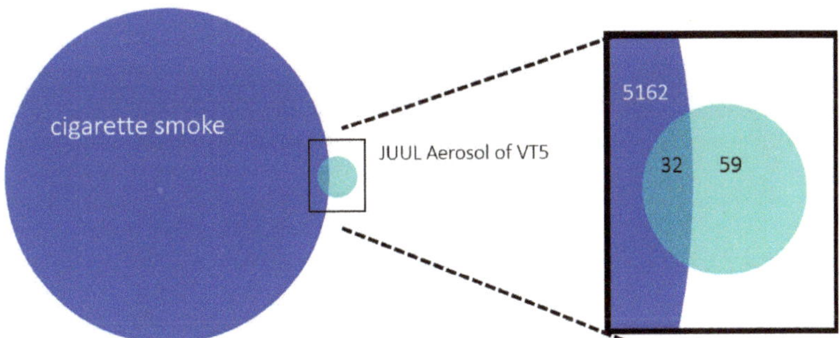

Figure 6. Comparison of the Compounds Detected in VT5 Aerosol to Compounds Found in Smoke of Combustible Cigarette (Circles are to Scale).

Table 12. Compounds Exclusive to JUUL VT5 Aerosol Compared to Cigarette Smoke.

Group Name	Intense Number of Compounds	Average Mass (µg/g) from Higher of Intense of Non-Intense	Average % Aerosol Mass
Flavorants	1	423.1	423.1×10^{-4}
HPHCs	ND	ND	ND
Extractables and Leachables	2	36.6	36.6×10^{-4}
Reaction Products	47	823.4	823.4×10^{-4}
Not Rationalized	9	38.7	38.7×10^{-4}
	59	1321.8	1321.8×10^{-4}

4. Discussion

Evaluating ENDS aerosols based only on knowledge of the harmful chemicals found in cigarette smoke leaves a gap in the assessment of ENDS aerosols [21,67]. This is due in part to the difference in regulated electronic heating and combustion as well as the ingredients unique to ENDS (i.e., propylene glycol and glycerol). Nicotine delivery from a CC is based on the combustion of plant material, whereas nicotine delivery from ENDS products is based on heating and aerosolizing nicotine-containing liquids. Measurement shows that JUUL operates within a regulated temperature range [68] with the intent of minimizing HPHCs formed as heat degradation byproducts [19,20,69]. Due to these fundamental differences in design, the use of only targeted methods to analyze ENDS products may leave a gap in our understanding of their aerosols. Therefore, in order to identify a wider range of the constituents contained within JUUL Virginia Tobacco aerosols, two complementary NTA methods were applied.

The LC–HRMS and GC–MS NTA methods presented here were optimized to screen for chemicals with a broad range of physiochemical properties, potentially present in ENDS aerosol. However, these methods are not exhaustive, and some chemical classes may not be well suited for a general sample collection and mass spectral analyses. More generally, mass spectrometry detection has limitations in its inability to detect nonionizable compounds, and compounds outside the defined mass-to-charge scan range. Another limitation inherent to a non-targeted analysis is the challenge in estimating the amounts of very high and very low concentration analytes in the same analysis; these NTA methods were developed to be sensitive to sub ppm range and were unable to provide reliable amount estimations for the detected primary constituents (PG, VG, nicotine, and benzoic acid), owing to the broad peak shape and detector saturation caused by their high concentrations. This, along with water, which was not detectable by these NTA methods, affected the ability to perform a careful mass balance analysis. Despite these limitations, the reported NTA approach provides a broad coverage of chemical properties—volatile, semi-volatile, non-volatile, non-polar, and polar compounds. Therefore, only a small percentage of aerosol constituents are thought to have gone undetected.

Overall, the NTA identified 88 and 91 chemical constituents (excluding PG, VG, nicotine, and benzoic acid) in VT3 and VT5, respectively. Of these compounds, 47% were confirmed using reference standards. Most of the compounds, approximately 50% of the NTA detected aerosol mass, were flavorants. Reaction products, including nicotine-related degradants, made up approximately 30–40% of the aerosol mass detected by NTA with the small amount remaining consisting of E&L and non-rationalized compounds. Overall, the 5 groups accounted for approximately 0.2-0.3% of the total aerosol mass with flavorants and reaction products comprising most of the mass detected. The unknown/not rationalized compounds were estimated to be present in low amounts and were detected with low signals, which posed challenges in compound identification. In total, more compounds were present in the 5.0% nicotine concentration product than the 3.0% nicotine product and more compounds were identified in aerosols generated with the intense puffing regimen than the non-intense puffing regimen. The NTA did not detect any HPHCs in the aerosol of JUUL Virginia Tobacco products.

The consistent composition of the aerosols across nicotine strengths and puffing conditions is contrasted with the variability in the composition of natural tobacco in traditional cigarettes. Botanical preparations contain a wide variety of bioactive secondary metabolites, which vary considerably depending on the cultivar and geography [70]. The variability of tobacco and the temperature of combustion (smoldering side stream smoke vs. mainstream smoke generated during the draw) both affect the composition of CC smoke, which is much more complex overall than the aerosol of ENDS products [11,71]. In contrast to CC, JUUL devices only produce aerosol during the draw and their electrical performance is well regulated [19,72]. This means that the puffing regimen should not appreciably impact the composition of the aerosol, which was shown to be true in this study, making a detailed and reproducible characterization of the chemicals possible.

In addition to the identification and semi-quantitation of the aerosol constituents, potential mechanisms for the formation of all tentatively identified compounds were considered. If an automated spectra search returned a tentative identification for which there was not a reasonable chemical mechanism of compound generation, then those identifications were re-evaluated. Understanding of the formation of the compounds detected by the non-targeted screening methods would have allowed for strategic formulation adjustment had that been necessary.

A global compilation of the 5162 compounds in CC smoke was catalogued by Rodgman and Perfetti [10], 93 of which have been identified by FDA as HPHCs. In addition to the specific links that are known between HPHCs and tobacco related diseases, there are additional risks related to the chemical complexity of CC smoke. Such chemical complexity has made it difficult to determine the active constituents responsible for all tobacco-related health risks of smoking and it is now being realized that a component of the health effects of this complex mixture are likely to result from a combined effect of these chemicals through multiple mechanisms rather than as a result of the effects of a single smoke constituent [9,71]. In this light, chemical complexity in and of itself may contribute to the harmfulness of cigarette smoke. Therefore, understanding the chemical complexity of JUUL aerosol in relation to CC smoke may aid in determining the relative potential health risks of using JUUL as an alternative to smoking for smokers who have not yet quit. The comparison of aerosol constituents detected by NTA to the list of chemicals in cigarette smoke catalogued by Perfetti and Rodgman resulted in 59 unique compound identifications in both VT3 and VT5 aerosols (out of 88 and 91 total constituents detected, respectively). Most of the aerosol mass from the 59 unique compounds in Virginia Tobacco products was comprised of one flavorant compound, triethyl citrate. Approximately 30% of the reaction products unique to VT aerosol were classified as nicotine-related compounds with limited structural information. Overall, the JUUL Virginia Tobacco aerosols studied here are shown to be approximately 50-fold less complex when compared to cigarette smoke.

The present study sought to construct a more complete appraisal of the full chemical space of JUUL Virginia Tobacco aerosol as a compliment to targeted analyses of pre-defined constituents, and to provide data for the comparative risk assessment of JUUL aerosols compared to CC smoking. The compound identifications and concentration estimates obtained by NTA provided a reasonably comprehensive characterization of JUUL Virginia Tobacco aerosols and proposed potential sources and chemical reactions for each compound allowed for a better understanding of aerosol composition and potential ingredient degradants. To this end, the present study contributes important understanding of the chemical composition of JUUL Virginia Tobacco aerosols toward appropriate assessment of the comparative public health risk for JUUL products compared to CC.

Supplementary Materials: The following are available online at https://www.mdpi.com/article/10.3390/separations8090130/s1, Table S1: Liquid Chromatographic Conditions; Table S2: Q-Exactive Mass Spectrometer Parameters; Table S3: Gas Chromatographic Parameters; Table S4: GC–MS Characterization of Aerosol Collected Under the Intense Puffing Regimen from Three Batches of Virginia Tobacco 3.0%; Table S5: GC–MS Characterization of Aerosol Collected Under the Non-Intense Puffing Regimen from Three Batches of Virginia Tobacco 3.0%; Table S6: LC–MS Characterization of Aerosol Collected Under the Intense Puffing Regimen from Three Batches of Virginia Tobacco 3.0%; Table S7: LC–HRMS Characterization of Aerosol Collected Under the Non-Intense Puffing Regimen from Three Batches of Virginia Tobacco 3.0%; Table S8: GC–MS Characterization of Aerosol Collected Under the Intense Puffing Regimen from Three Batches of Virginia Tobacco 5.0%; Table S9: GC–MS Characterization of Aerosol Collected Under the Non-Intense Puffing Regimen from Three Batches of Virginia Tobacco 5.0%; Table S10: LC–HRMS Characterization of Aerosol Collected Under the Intense Puffing Regimen from Three Batches of Virginia Tobacco 5.0%; Table S11: LC–HRMS Characterization of Aerosol Collected Under the Non-Intense Puffing Regimen from Three Batches of Virginia Tobacco 5.0%; Table S12: List of Compounds Detected in the Aerosol Collected from JUUL Virginia Tobacco 3.0% under the Intense and Non-Intense Puffing Regimens by Non-Targeted Analyses; Table S13: List of Compounds Detected in the Aerosol Collected from JUUL Virginia Tobacco 5.0% Under the Intense and Non-Intense Puffing Regimens by Non-Targeted Analyses.

Author Contributions: M.R.C.: conceptualization, data curation, investigation, formal analysis, methodology, and writing; L.N.J.: conceptualization, data curation, investigation, methodology, and writing; P.C.B.: visualization and writing; A.L.: data curation and conceptualization; C.Y.: software; A.O.: project administration; J.B.J.: supported analysis; I.G.G.: supervision and resources. All authors have read and agreed to the published version of the manuscript.

Funding: This research received no external funding.

Institutional Review Board Statement: Not applicable.

Informed Consent Statement: Not applicable.

Data Availability Statement: Due to its proprietary nature, the analyzed data is not publicly available.

Conflicts of Interest: The authors declare no conflict of interest.

References

1. US Department of Health and Human Services. *The Health Consequences of Smoking—50 Years of Progress: A Report of the Surgeon General*; Reports of the Surgeon General: Altanta, GA, USA, 2014.
2. *Report on the Global Tobacco Epidemic: A Report of the World Health Organization*; World Health Organization: Geneva, Switzerland, 2019.
3. FDA. FDA's Youth Tobacco Prevention Plan. Available online: https://www.fda.gov/tobacco-products/youth-and-tobacco/fdas-youth-tobacco-prevention-plan (accessed on 18 August 2021).
4. USSG. *Preventing Tobacco Use Among Youth and Young Adults: A Report of the Surgeon General*; National Center for Chronic Disease Prevention and Health Promotion (US) Office on Smoking and Health: Atlanta, GA, USA, 2012.
5. Family Smoking Prevention and Tobacco Control and Federal Retirement Reform. Available online: https://www.govinfo.gov/content/pkg/PLAW-111publ31/pdf/PLAW-111publ31.pdf (accessed on 18 August 2021).
6. Brown, J.; Brown, B.; Schwiebert, P.; Ramakrisnan, K.; McCarthy, L.H. In adult smokers unwilling or unable to quit, does changing from tobacco cigarettes to electronic cigarettes decrease the incidence of negative health effects associated with smoking tobacco? A Clin-IQ. *J. Patient Cent. Res. Rev.* **2014**, *1*, 99–101. [CrossRef]
7. US Food and Drug Administration. FDA's Comprehensive Plan for Tobacco and Nicotine Regulation. 2019. Available online: https://www.fda.gov/tobacco-products/ctp-newsroom/fdas-comprehensive-plan-tobacco-and-nicotine-regulation (accessed on 18 August 2021).
8. Benowitz, N.L. Nicotine addiction. *N. Engl. J. Med.* **2010**, *362*, 2295–2303. [CrossRef] [PubMed]
9. Paschke, M.; Hutzler, C.; Henkler, F.; Luch, A. Oxidative and inert pyrolysis on-line coupled to gas chromatography with mass spectrometric detection: On the pyrolysis products of tobacco additives. *Int. J. Hyg. Environ. Health* **2016**, *219*, 780–791. [CrossRef] [PubMed]
10. Rodgman, A.; Perfetti, T.A. *The Chemical Components of Tobacco and Tobacco Smoke*, 2nd ed.; CRC Press: Boca Raton, FL, USA, 2013.
11. *How Tobacco Smoke Causes Disease: The Biology and Behavioral Basis for Smoking-Attributable Disease: A Report of the Surgeon General*; Publications and Reports of the Surgeon General: Atlanta, GA, USA, 2010.
12. IARC Working Group on the Evaluation of Carcinogenic Risks to Humans; World Health Organization; International Agency for Research on Cancer. *Tobacco smoke and involuntary smoking. IARC Monogr. Eval. Carcinog. Risks Hum.* **2004**, *83*, 1–1438.
13. Burns, D.M.; Dybing, E.; Gray, N.; Hecht, S.; Anderson, C.; Sanner, T.; O'Connor, R.; Djordjevic, M.; Dresler, C.; Hainaut, P.; et al. Mandated lowering of toxicants in cigarette smoke: A description of the World Health Organization TobReg proposal. *Tob. Control* **2008**, *17*, 132–141. [CrossRef] [PubMed]
14. FDA. Harmful and Potentially Harmful Constituents in Tobacco Products; Established List; Proposed Additions; Request for Comments. *Fed Regist* **2019**, *2021*, 38032–38035.
15. Dusautoir, R.; Zarcone, G.; Verriele, M.; Garçon, G.; Fronval, I.; Beauval, N.; Allorge, D.; Riffault, V.; Locoge, N.; Lo-Guidice, J.-M.; et al. Comparison of the chemical composition of aerosols from heated tobacco products, electronic cigarettes and tobacco cigarettes and their toxic impacts on the human bronchial epithelial BEAS-2B cells. *J. Hazard. Mater.* **2021**, *401*, 123417. [CrossRef]
16. Tayyarah, R.; Long, G.A. Comparison of select analytes in aerosol from e-cigarettes with smoke from conventional cigarettes and with ambient air. *Regul. Toxicol. Pharm.* **2014**, *70*, 704–710. [CrossRef] [PubMed]
17. McNeill, A.; Brose, L.S.; Calder, R.; Bauld, L.; Robson, D. *Evidence Review of E-Cigarettes and Heated Tobacco Products 2018*; A Report Comissioned by Public Health England: London, UK, 2018.
18. Gillman, I.G.; Kistler, K.A.; Stewart, E.W.; Paolantonio, A.R. Effect of variable power levels on the yield of total aerosol mass and formation of aldehydes in e-cigarette aerosols. *Regul. Toxicol. Pharm.* **2016**, *75*, 58–65. [CrossRef]

19. GIllman, I.G.; Johnson, M.; Martin, A.; Meyers, D.; Alston, B.; Misra, M. Characterization of Temperature Regulation and HPHC Profile of a Nicotine-Salt Based ENDS Product. In Proceedings of the Society for Research on Nicotine, Baltimore, MD, USA, 21–24 February 2018.

20. Farsalinos, K.E.; Kistler, K.A.; Pennington, A.; Spyrou, A.; Kouretas, D.; Gillman, I.G. Aldehyde levels in e-cigarette aerosol: Findings from a replication study and from use of a new-generation device. *Food Chem. Toxicol.* **2018**, *111*, 64–70. [CrossRef] [PubMed]

21. El-Hage, R.; El-Hellani, A.; Salman, R.; Talih, S.; Shihadeh, A.; Saliba, N.A. Vaped Humectants in E-Cigarettes Are a Source of Phenols. *Chem. Res. Toxicol.* **2020**, *33*, 2374–2380. [CrossRef]

22. Wilson, S.; Partos, T.; McNeill, A.; Brose, L.S. Harm perceptions of e-cigarettes and other nicotine products in a UK sample. *Addiction* **2019**, *114*, 879–888. [CrossRef]

23. Caponnetto, P. Well-being and harm reduction, the consolidated reality of electronic cigarettes ten years later from this emerging phenomenon: A narrative review. *Health Psychol. Res.* **2021**, *8*, 9463. [CrossRef]

24. Bentley, M.C.; Almstetter, M.; Arndt, D.; Knorr, A.; Martin, E.; Pospisil, P.; Maeder, S. Comprehensive chemical characterization of the aerosol generated by a heated tobacco product by untargeted screening. *Anal. Bioanal. Chem.* **2020**, *412*, 2675–2685. [CrossRef]

25. Woelfel, K.; Hartman, T.G. Mass Spectrometry of the Acetal Derivatives of Selected Generally Recognized as Safe Listed Aldehydes with Ethanol, 1,2-Propylene Glycol and Glycerol. In *Flavor Analysis*; ACS Symposium Series; American Chemical Society: Washintong, DC, USA, 1998; Volume 705, pp. 193–210.

26. Erythropel, H.C.; Davis, L.M.; de Winter, T.M.; Jordt, S.E.; Anastas, P.T.; O'Malley, S.S.; Krishnan-Sarin, S.; Zimmerman, J.B. Flavorant-Solvent Reaction Products and Menthol in JUUL E-Cigarettes and Aerosol. *Am. J. Prev. Med.* **2019**, *57*, 425–427. [CrossRef]

27. Jabba, S.V.; Diaz, A.N.; Caceres, A.I.; Erythropel, H.C.; Kumar, V.; Varghese, S.; Simmerman, J.B.; Jordt, S.E. Chemical Adducts of Reactive Flavor Aldehydes Formed in E-cigarette Liquids Are Cytotoxic and Inhibit Mitochondrial Function in Respiratory Epithelial Cells. In Proceedings of the Society of Toxicology, Online, 12–26 March 2021.

28. Erythropel, H.C.; Jabba, S.V.; DeWinter, T.M.; Mendizabal, M.; Anastas, P.T.; Jordt, S.E.; Zimmerman, J.B. Formation of flavorant-propylene Glycol Adducts With Novel Toxicological Properties in Chemically Unstable E-Cigarette Liquids. *Nicotine Tob. Res.* **2019**, *21*, 1248–1258. [CrossRef] [PubMed]

29. Ward, A.M.; Yaman, R.; Ebbert, J.O. Electronic nicotine delivery system design and aerosol toxicants: A systematic review. *PLoS ONE* **2020**, *15*, e0234189. [CrossRef] [PubMed]

30. Vreeke, S.; Korzun, T.; Luo, W.; Jensen, R.P.; Peyton, D.H.; Strongin, R.M. Dihydroxyacetone levels in electronic cigarettes: Wick temperature and toxin formation. *Aerosol. Sci. Technol.* **2018**, *52*, 370–376. [CrossRef] [PubMed]

31. Bitzer, Z.T.; Goel, R.; Reilly, S.M.; Elias, R.J.; Silakov, A.; Foulds, J.; Muscat, J.; Richie, J.P., Jr. Effect of flavoring chemicals on free radical formation in electronic cigarette aerosols. *Free Radic. Biol. Med.* **2018**, *120*, 72–79. [CrossRef] [PubMed]

32. Kosmider, L.; Spindle, T.R.; Gawron, M.; Sobczak, A.; Goniewicz, M.L. Nicotine emissions from electronic cigarettes: Individual and interactive effects of propylene glycol to vegetable glycerin composition and device power output. *Food Chem. Toxicol.* **2018**, *115*, 302–305. [CrossRef] [PubMed]

33. Zhao, J.; Zhang, Y.; Sisler, J.D.; Shaffer, J.; Leonard, S.S.; Morris, A.M.; Qian, Y.; Bello, D.; Demokritou, P. Assessment of reactive oxygen species generated by electronic cigarettes using acellular and cellular approaches. *J. Hazard. Mater.* **2018**, *344*, 549–557. [CrossRef]

34. Farsalinos, K.E.; Gillman, I.G. Carbonyl Emissions in E-cigarette Aerosol: A Systematic Review and Methodological Considerations. *Front. Physiol.* **2017**, *8*, 1119. [CrossRef]

35. Klupinski, T.P.; Strozier, E.D.; Friedenberg, D.A.; Brinkman, M.C.; Gordon, S.M.; Clark, P.I. Identification of New and Distinctive Exposures from Little Cigars. *Chem. Res. Toxicol.* **2016**, *29*, 162–168. [CrossRef] [PubMed]

36. Arndt, D.; Wachsmuth, C.; Buchholz, C.; Bentley, M. A complex matrix characterization approach, applied to cigarette smoke, that integrates multiple analytical methods and compound identification strategies for non-targeted liquid chromatography with high-resolution mass spectrometry. *Rapid Commun. Mass Spectrom.* **2020**, *34*, e8571. [CrossRef] [PubMed]

37. Klupinski, T.P.; Strozier, E.D.; Makselan, S.D.; Buehler, S.S.; Peters, E.N.; Lucas, E.A.; Casbohm, J.S.; Friedenberg, D.A.; Landgraf, A.J.; Frank, A.J.; et al. Chemical characterization of marijuana blunt smoke by non-targeted chemical analysis. *Inhal. Toxicol.* **2020**, *32*, 177–187. [CrossRef] [PubMed]

38. Holt, A.K.; Poklis, J.L.; Cobb, C.O.; Peace, M.R. The Identification of Gamma-Butyrolactone in JUUL Liquids. *J. Anal. Toxicol.* **2021**. [CrossRef] [PubMed]

39. Miller, J.H.; Shah, N.H.; Noe, M.R.; Agnew-Heard, K.A.; Gardner, W.P.; Pithawalla, Y.P. Non-targeted analysis using gas chromatography mass spectrometry for evaluation of chemical composition of e-vapor products. In Proceedings of the CORESTA, Online, 12 October–12 November 2020.

40. Rawlinson, C.; Martin, S.; Frosina, J.; Wright, C. Chemical characterisation of aerosols emitted by electronic cigarettes using thermal desorption–gas chromatography–time of flight mass spectrometry. *J. Chromatogr. A* **2017**, *1497*, 144–154. [CrossRef] [PubMed]

41. Herrington, J.S.; Myers, C. Electronic cigarette solutions and resultant aerosol profiles. *J. Chromatogr. A* **2015**, *1418*, 192–199. [CrossRef] [PubMed]

42. Xiao, Z.; Yu, D.; Niu, Y.; Chen, F.; Song, S.; Zhu, J.; Zhu, G. Characterization of aroma compounds of Chinese famous liquors by gas chromatography-mass spectrometry and flash GC electronic-nose. *J. Chromatogr. B Anal. Technol. Biomed. Life Sci.* **2014**, *945–946*, 92–100. [CrossRef]

43. Mehari, B.; Redi-Abshiro, M.; Chandravanshi, B.S.; Combrinck, S.; McCrindle, R.; Atlabachew, M. GC-MS profiling of fatty acids in green coffee (Coffea arabica L.) beans and chemometric modeling for tracing geographical origins from Ethiopia. *J. Sci. Food Agric.* **2019**, *99*, 3811–3823. [CrossRef]

44. Zhang, H.; Li, Y.; Mi, J.; Zhang, M.; Wang, Y.; Jiang, Z.; Hu, P. GC-MS Profiling of Volatile Components in Different Fermentation Products of Cordyceps Sinensis Mycelia. *Molecules* **2017**, *22*, 1800. [CrossRef] [PubMed]

45. Wang, Y.; Zhao, J.; Xu, F.; Wu, X.; Hu, W.; Chang, Y.; Zhang, L.; Chen, J.; Liu, C. GC-MS, GC-O and OAV analyses of key aroma compounds in Jiaozi Steamed Bread. *Grain Oil Sci. Technol.* **2020**, *3*, 9–17. [CrossRef]

46. Stupak, M.; Goodall, I.; Tomaniova, M.; Pulkrabova, J.; Hajslova, J. A novel approach to assess the quality and authenticity of Scotch Whisky based on gas chromatography coupled to high resolution mass spectrometry. *Anal. Chim. Acta* **2018**, *1042*, 60–70. [CrossRef]

47. Angioni, A.; Pintore, G.A.; Caboni, P. Determination of wine aroma compounds by dehydration followed by GC/MS. *J. AOAC Int.* **2012**, *95*, 813–819. [CrossRef] [PubMed]

48. Krüsemann, E.J.Z.; Visser, W.F.; Cremers, J.W.J.M.; Pennings, J.L.A.; Talhout, R. Identification of flavour additives in tobacco products to develop a flavour library. *Tob. Control* **2018**, *27*, 105. [CrossRef] [PubMed]

49. Tahri, K.; Duarte, A.A.; Carvalho, G.; Ribeiro, P.A.; da Silva, M.G.; Mendes, D.; El Bari, N.; Raposo, M.; Bouchikhi, B. Distinguishment, identification and aroma compound quantification of Portuguese olive oils based on physicochemical attributes, HS-GC/MS analysis and voltammetric electronic tongue. *J. Sci. Food Agric.* **2018**, *98*, 681–690. [CrossRef] [PubMed]

50. Pagano, T.; DiFrancesco, A.G.; Smith, S.B.; George, J.; Wink, G.; Rahman, I.; Robinson, R.J. Determination of Nicotine Content and Delivery in Disposable Electronic Cigarettes Available in the United States by Gas Chromatography-Mass Spectrometry. *Nicotine Tob. Res.* **2015**, *18*, 700–707. [CrossRef] [PubMed]

51. Available online: https://www.nist.gov/programs-projects/nist20-updates-nist-tandem-and-electron-ionization-spectral-libraries (accessed on 18 August 2021).

52. Crosswhite, M.; Jeong, L.; Yang, C.; Jameson, B.; Oldham, M.; Cook, D.; Gillman, I.G. Comprehensive Evaluation of Aerosol Constituents from JUUL Virginia Tobacco 5.0% Using Non-Targeted Analysis: LC-HRMS Analysis of E-Vapor Aerosol, Data Acquisition and Processing. In Proceedings of the North Carolina Local Section of the American Chemical Society, Online, 10 November 2020.

53. Sobus, J.R.; Grossman, J.N.; Chao, A.; Singh, R.; Williams, A.J.; Grulke, C.M.; Richard, A.M.; Newton, S.R.; McEachran, A.D.; Ulrich, E.M. Using prepared mixtures of ToxCast chemicals to evaluate non-targeted analysis (NTA) method performance. *Anal. Bioanal. Chem.* **2019**, *411*, 835–851. [CrossRef] [PubMed]

54. Caesar, L.K.; Kvalheim, O.M.; Cech, N.B. Hierarchical cluster analysis of technical replicates to identify interferents in untargeted mass spectrometry metabolomics. *Anal. Chim. Acta* **2018**, *1021*, 69–77. [CrossRef] [PubMed]

55. Bruni, R.; Brighenti, V.; Caesar, L.K.; Bertelli, D.; Cech, N.B.; Pellati, F. Analytical methods for the study of bioactive compounds from medicinally used Echinacea species. *J. Pharm. Biomed. Anal.* **2018**, *160*, 443–477. [CrossRef] [PubMed]

56. Flavor Ingredient Library. Available online: https://www.femaflavor.org/flavor-library (accessed on 28 April 2021).

57. Jenke, D.; Carlson, T. A Compilation of Safety Impact Information for Extractables Associated with Materials Used in Pharmaceutical Packaging, Delivery, Administration, and Manufacturing Systems. *PDA J. Pharm. Sci. Technol.* **2014**, *68*, 407–455. [CrossRef] [PubMed]

58. Jenke, D. Appendix: Materials Used in Pharmaceutical Constructs and their Associated Extractables. In *Compatibility of Pharmaceutical Products and Contact Materials*; John Wiley & Sons, Inc.: Hoboken, NJ, USA, 2009; pp. 347–370.

59. Cederberg, T.; Jensen, L.K. *Siloxanes in Silicone Products Intended for Food Contact: Selected Samples from the Norwegian Market in 2016*; National Food Institute, Technical University of Denmark: Copenhagen, Denmark, 2017.

60. *Vapour Products—Routine Analytical Vaping Machine—Definitions and Standard Conditions*; International Organization for Standardization 20768:2018(E); ISO: Geneva, Switzerland, 2018.

61. Belushkin, M.; Tafin Djoko, D.; Esposito, M.; Korneliou, A.; Jeannet, C.; Lazzerini, M.; Jaccard, G. Selected Harmful and Potentially Harmful Constituents Levels in Commercial e-Cigarettes. *Chem. Res. Toxicol.* **2020**, *33*, 657–668. [CrossRef] [PubMed]

62. Statistical Methods for Research Workers. *J. R. Stat. Soc.* **1926**, *89*, 144–145. [CrossRef]

63. Fraley, N.; Anderson, E.; Almond, C. Glycidol in Tobacco Heated Products. In Proceedings of the TSRC, Leesburg, VA, USA, 15–18 September 2019.

64. Gilman, G.; Johnson, M.; Martin, A.; Misra, M. HPHC Analysis of Eight Flavors of a Temperature-Regulated Nicotine Salt-Based ENDS Product. In Proceedings of the 5th Global Forum on Nicotine, Warsaw, Poland, 14 June 2018.

65. 65. RJ Reynolds Tobacco Co. *Chemical and Biological Studies on New Cigarette Prototypes that Heat Instead of Burn Tobacco*; RJ Reynolds Tobacco Company: Winston-Salem, NC, USA, 1988; pp. 144–175.

66. Baker, R.R.; Bishop, L.J. The pyrolysis of tobacco ingredients. *J. Anal. Appl. Pyrolysis* **2004**, *71*, 223–311. [CrossRef]

67. King, B.; Borland, R.; Morphett, K.; Gartner, C.; Fielding, K.; O'Connor, R.J.; Romijnders, K.; Talhout, R. 'It's all the other stuff!' How smokers understand (and misunderstand) chemicals in cigarettes and cigarette smoke. *Public Underst. Sci.* **2021**, *6*, 777–796. [CrossRef]

68. Alston, B. Measurement of temperature regulation performance of the JUUL nicotine salt pod system. In Proceedings of the CORESTA, Hamburg, Germany, 6–10 October 2019.
69. Salamanca, J.C.; Meehan-Atrash, J.; Vreeke, S.; Escobedo, J.O.; Peyton, D.H.; Strongin, R.M. E-cigarettes can emit formaldehyde at high levels under conditions that have been reported to be non-averse to users. *Sci. Rep.* **2018**, *8*, 7559. [CrossRef] [PubMed]
70. Kellogg, J.J.; Graf, T.N.; Paine, M.F.; McCune, J.S.; Kvalheim, O.M.; Oberlies, N.H.; Cech, N.B. Comparison of Metabolomics Approaches for Evaluating the Variability of Complex Botanical Preparations: Green Tea (Camellia sinensis) as a Case Study. *J. Nat. Prod.* **2017**, *80*, 1457–1466. [CrossRef] [PubMed]
71. Saha, S.P.; Bhalla, D.K.; Whayne, T.F., Jr.; Gairola, C. Cigarette smoke and adverse health effects: An overview of research trends and future needs. *Int. J. Angiol.* **2007**, *16*, 77–83. [CrossRef]
72. Talih, S.; Salman, R.; El-Hage, R.; Karam, E.; Salam, S.; Karaoghlanian, N.; El-Hellani, A.; Saliba, N.; Shihadeh, A. A comparison of the electrical characteristics, liquid composition, and toxicant emissions of JUUL USA and JUUL UK e-cigarettes. *Sci. Rep.* **2020**, *10*, 7322. [CrossRef] [PubMed]

 *separations*

Article

1,2-Propylene Glycol: A Biomarker of Exposure Specific to e-Cigarette Consumption

Therese Burkhardt, Nikola Pluym, Gerhard Scherer and Max Scherer *

ABF Analytisch-Biologisches Forschungslabor GmbH, Semmelweisstr. 5, 82152 Planegg, Germany;
therese.burkhardt@abf-lab.com (T.B.); nikola.pluym@abf-lab.com (N.P.); gerhard.scherer@abf-lab.com (G.S.)
* Correspondence: max.scherer@abf-lab.com

Abstract: Over the past decade, new emerging tobacco and nicotine-delivery products have changed the tobacco landscape. Especially, electronic cigarettes (ECs) have been suggested to be considered for tobacco harm reduction, reinforcing the need to identify novel biomarkers of exposure (BoE) specific to the EC use as this would complement exposure assessment and product compliance monitoring. Therefore, a sensitive LC-MS/MS method for the quantification of 1,2-propylene glycol (PG) and glycerol (G), the main e-liquid constituents, was established. PG and G were analyzed in plasma and urine samples from a clinical study comparing five nicotine product user groups, users of combustible cigarettes (CC), electronic cigarettes (EC), heated tobacco products (HTP), oral tobacco (OT), and oral/dermal nicotine delivery products (used for nicotine replacement therapy, NRT) with a control group of non-users (NU). Data demonstrate significantly elevated PG levels in urine and plasma in EC users compared to users of CC, HTP, NRT, OT as well as NU. In addition, PG in plasma and urine of vapers significantly correlated with nicotine (plasma) and total nicotine equivalents (urine), biomarkers reflecting product consumption, emphasizing the high specificity of PG as a BoE for EC consumption. We therefore suggest the use of PG as BoE in urine and/or plasma in order to monitor EC use compliance in exposure assessments.

Keywords: propylene glycol; electronic cigarette; biomarker of exposure; compliance marker

Citation: Burkhardt, T.; Pluym, N.; Scherer, G.; Scherer, M. 1,2-Propylene Glycol: A Biomarker of Exposure Specific to e-Cigarette Consumption. *Separations* **2021**, *8*, 180. https://doi.org/10.3390/separations8100180

Academic Editor: Fadi Aldeek

Received: 13 August 2021
Accepted: 2 October 2021
Published: 9 October 2021

Publisher's Note: MDPI stays neutral with regard to jurisdictional claims in published maps and institutional affiliations.

1. Introduction

Over decades, the measurement of biomarkers of exposure (BoE) has contributed important data to evaluate the health risk from cigarette smoking [1]. The two most common BoE that can be evaluated for all nicotine containing products are nicotine itself, as the most abundant alkaloid found in the tobacco leaf [2,3] as well as cotinine, its major metabolite possessing a longer half-life. Other relevant urinary biomarkers in tobacco smoke exposure assessment originate from tobacco-specific nitrosamines (TSNAs) (e.g., NNN, NNAL), polycyclic aromatic hydrocarbons (PAHs) (e.g., 1-hydroxypyrene, 3-hydroxybenzo[a]pyrene), aromatic amines (e.g., ortho-toluidine, 1-/2-naphthylamine, 3-/4-aminobiphenyl), and mercapturic acids of volatile organic compounds (VOCs) (e.g., 3-HPMA, CEMA) [2,4–9]. In the past decade, the tobacco landscape has changed, and a variety of new tobacco and nicotine-delivery products have been developed that pose a potentially reduced risk for the consumer as compared to smoking cigarettes. In 2015, Public Health England suggested to consider electronic cigarettes (ECs) for tobacco harm reduction, as complete switching could help reduce smoking related diseases [10]. They substantiated this claim in their most recent evidence update report in 2021 to which vaping of ECs is positively associated with successfully quitting smoking [11]. Still, a controversial debate about the benefits and risks of ECs continues to date [12]. Obviously, there is a need to identify BoE specific to EC consumption for a profound exposure and risk assessment [13–16]. However, to our knowledge, there is as yet no specific BoE to distinguish the use of ECs from the concomitant use of other tobacco/ nicotine products (dual or multiple product use).

E-liquids of ECs contain, in addition to nicotine, flavoring chemicals and carrier solvents, which are often referred to as humectants or stabilizing agents [17]. Mainly 1,2-propylene glycol (PG) and glycerol (G) are used as carrier solvents, constituting 80–95% of the e-liquid [17–19]. The vaporized PG and G generate an aerosol which is inhaled and thus absorbed by the user. Up to 45% of the unchanged PG and G are excreted in urine [20–22], thus becoming potential biomarker candidates for EC consumption. Schick et al. already discussed PG as appropriate BoE for EC consumption, but considered it as not suitable due to its widespread occurrence in daily use consumer products [5]. PG and G, as color- and odorless, water-soluble fluids, have beneficial properties as solvents, humectants, and antifreeze agents, making it attractive for a variety of applications across various industries such as food, pharmaceutical, cosmetic, medical, the manufacture of paints and coatings, and the production of plasticizers and polyester resins [17,23–26]. The widespread use of PG and G resulting in a general exposure of the population requires to verify, whether significant differences in the PG and G exposure are detectable between EC users (vapers) and users of other tobacco and nicotine-containing products as well as non-users [5]. The differentiation between user groups of different tobacco/nicotine products could provide a better understanding of the exposure pattern and the related health effects. In addition, BoE or BoE patterns specific for the use of an individual product, such as EC, would be useful to monitor product compliance (ideally the sole use of one product) rather than relying on self-reports, which is of particular importance for epidemiological studies [4].

For this purpose, a controlled clinical study was conducted comparing five tobacco/nicotine product user groups, namely, smokers of combustible cigarettes (CC), EC vapers, heated tobacco product (HTP) users, oral tobacco (OT) users, and users of oral/dermal nicotine delivery products (used for nicotine replacement therapy, NRT) as well as a control group of non-users (NU) [27]. Urine and plasma samples were analyzed for their PG and G content and statistical evaluation was performed across the different product user groups. Furthermore, PG and G levels were investigated for their association with the vaping intensity.

2. Materials and Methods

2.1. Study Design

A controlled, single-center, open label trial was conducted comparing five nicotine product user groups, namely exclusive users of CC, EC, HTP, OT, NRT with a control group of non-users (NU). Detailed information regarding the study design and the study population is described in Sibul et al. [27]. The study protocol has been approved by the ethics committee of the Medical Association Hamburg. Ten subjects per group were confined for 76 h (diet-control, exclusive use of one product), during which free, uncontrolled use of the products (own brand) was allowed. The amount of PG and G in the e-liquids consumed by the subjects ranged from of 50–55% for PG and 45–50% for G, respectively. This information is based on the manufacturer's specifications of the consumed liquids, or on the self-reported PG/G content of the e-liquid base if the liquids were self-mixed. Blood samples were collected at 7 a.m. and 5 p.m. on each day starting in the evening of day -1, when the subjects were admitted to the clinic. In total, 420 plasma samples were analyzed in this study. All urine voids were collected separately throughout the course of the clinical study. The total volume of each void was determined gravimetrically together with the time of void. Urine fractions were pooled to get 12 h urines (U0, U1/2, U3, U4/5, U6, U7/8; Figure 1). For PG/G analysis, six 12 h- urine pools of the 76-h stay of each subject were analyzed resulting in a total number of 360 urine samples.

B – blood draw (processed to receive plasma)
U – urine fraction

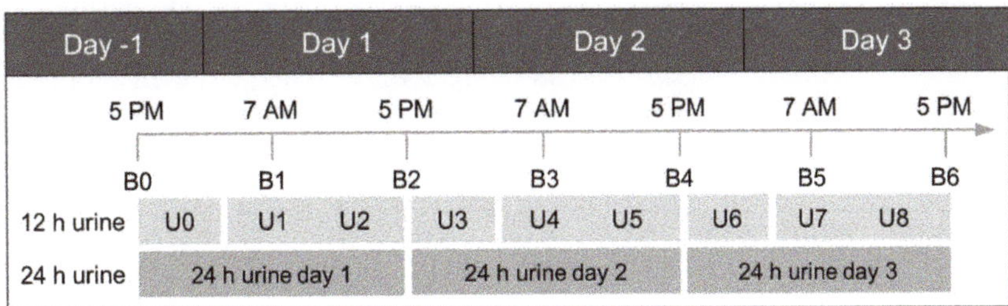

Figure 1. Time schedule for sample collection.

2.2. Reagents and Chemicals

Benzoyl chloride ($\geq$99%), glycine ($\geq$99%), and sodium hydroxide ($\geq$99%) were purchased from Merck KGaA (Darmstadt, Germany). n-Pentane ($\geq$99%) was supplied by VWR International GmbH (Darmstadt, Germany). Acetonitrile (min. 99.97%) was obtained from Th. Geyer GmbH & Co. KG (Renningen, Germany), bovine plasma from Biowest SAS (Nuaillé, France), and formic acid ($\geq$99%) from Biosolve (Dieuze, France).

Reference compounds 1,2-propylene glycol (99.9%) and glycerol (99.9%) were purchased from Sigma-Aldrich$^{®}$ a member of Merck KGaA (Darmstadt, Germany). Internal standards 1,2-propylene glycol-d_6 (99.6%) and glycerol-d_5 (98%) were obtained from CDN Isotopes Inc. (Quebec, QC, Canada).

2.3. Analytical Method

Urine and plasma samples were analyzed for their PG and G content according to Landmesser et al. [28]. An enzymatic hydrolysis experiment in urine using glucuronidase and sulfatase did not show an increase in PG and G concentration (data not shown), consequently this step was omitted for the sample preparation. In brief, 10 µL of an internal standard (IS) mixture containing 5 µg/mL 1,2-propylene glycol-d_6 and 5 µg/mL glycerol-d_5 were mixed with 25 µL of the urine or plasma sample. Derivatization was achieved by the addition of 500 µL of 4 M sodium hydroxide and 100 µL benzoyl chloride initiating the Schotten-Baumann reaction. n-Pentane (2 mL) was added and stirred for 15 min on a multi-tube vortex mixer. In order to quench the excess derivatization agent, 500 µL of a glycine solution in water (10% (v/v)) was added and subsequently mixed for another 15 min. Mixtures containing plasma were additionally precipitated at <-70 °C for 15 min. After centrifugation of the sample (10 min, 1860 rcf), the supernatant was transferred into a new tube and evaporated to dryness using a vacuum concentrator. The sample was reconstituted in 100 µL of acetonitrile and analyzed by LC-MS/MS.

Analysis was performed by using an HTC PAL autosampler (CTC Analytics AG, Zwingen, Switzerland) with an Agilent 1100 HPLC system (Agilent Technologies, Inc., Waldbronn, Germany) hyphenated to an API 4000^{TM} triple quadrupole mass spectrometer (MS/MS) (Sciex, Darmstadt, Germany). Analyst$^{®}$ Software (Version 1.5.3, Sciex, Framingham, United States) was used for data acquisition and quantification. A Kinetex$^{®}$ 5 µm EVO C18 (100 Å, 150 × 2.1 mm, Phenomenex Ltd., Aschaffenburg, Germany) equipped with a SecurityGuard^{TM} ULTRA cartridge system for EVO-C18 (ID 2.1 mm Phenomenex Ltd., Aschaffenburg, Germany) as pre-column was used for chromatographic separation with an injection volume of 10 µL. The mobile phase consisted of 0.1% formic acid in water (A) and acetonitrile (B). Flow rate was set to 1.0 mL/min applying gradient elution as follows: initial conditions of 50% B were held for 0.6 min, increased to 60% B until 0.7 min, held at 60% B for 1.3 min, further increased to 80% B over the next 0.1 min, held for 1.9 min, increased to 95% B over 0.01 min, held for 1.99 min, decreased to 50% B within 0.01 min and held at 50% B for 1.99 min for re-equilibration, resulting in a total runtime of 8 min. Oven temperature was maintained at 40 °C. MS acquisition was carried out with an electro spray ionization (ESI) ion source operated in positive ion mode. Source parameters were as follows: curtain gas: 30 psi, ion spray voltage: 5500 V, temperature: 300 °C, ion source gas 1: 60 psi, ion source gas 2: 50 psi, collision gas: 10 psi. The MS was operated in multiple reaction monitoring (MRM) mode with parameters specified in Table 1. Data were evaluated using Analyst® Software (Version 1.5.3, Sciex, Framingham, MA, USA) and Excel 2013 (Microsoft Cooperation, Redmond, WA, USA). Quadratic regression with $1/y$ weighting was applied.

Table 1. MS/MS parameter for 1,2-propylene glycol (PG) and glycerol (G).

Analyte	Q1 m/z (Da)	Q3 m/z (Da)		Declustering Potential (V)	Collision Energy (V)	Collision Cell Exit Potential (V)
1,2-Propylene glycol (PG)	285	163	Quantifier	86	13	10
	285	105	Qualifier	86	31	8
Glycerol (G)	405	283	Quantifier	91	13	16
	405	105	Qualifier	91	37	8
1,2-Propylene glycol-d_6	291	169	IS for PG	86	13	10
Glycerol-d_5	410	288	IS for G	91	13	16

The determination of PG and G was performed in separate batches consisting of unknown samples, QC samples at low, medium, and high levels, calibrators, and blanks. Quantification of the study samples was performed by using water as surrogate matrix, as no analyte-free plasma or urine samples were available. Each calibration consisted of a blank, a zero, and eight non-zero concentration levels, including the LLOQ (lowest calibrator). Calibration ranged from 0.1 to 150 µg/mL. Deviation from the target values were evaluated for accepting calibrators and to verify the calibration range. The LLOQ was 0.1 µg/mL for PG and G in urine and plasma, respectively, and determined during method validation with a signal-to-noise ratio of at least 9 under consideration of the background levels, an accuracy of 80–120%, and a precision of ±20%. Human spot urine samples and bovine plasma spiked with the analytes to achieve three different concentration levels (low (L), medium (M), high (H)) reflecting the expected concentration range of the study samples were prepared as quality controls (QCs). For plasma QCs, 50 µg/mL PG and G (L) and 250 µg/mL PG and G (M, H) in water were added to bovine plasma so that the final QC concentrations were 0.4 µg/mL (L), 11.9 µg/mL (M), 115.5 µg/mL (H) for PG and 5.0 µg/mL (L) 15.4 µg/mL (M), 95.9 µg/mL (H) for G, respectively. For urinary QCs, 5 µg/mL G (L, PG present natively), 100 µg/mL (M) and 250 µg/mL (H) PG and G in water, respectively, were added to human urine so that the final QC concentrations were 0.7 µg/mL (L), 10.1 µg/mL (M), 123.7 µg/mL (H) for PG and 0.6 µg/mL (L), 9.8 µg/mL (M), 119.2 µg/mL (H) for G, respectively. More than 5% of unknown samples per analytical run (or at least 6 QC samples, two QC samples for each level) were randomly interspersed across the analytical runs as QC samples covering the expected range of analyte concentrations. This results in a total number of 27 QCs (9 per level) for PG and 24 QCs (8 per level) for G in urine and 27 QCs (9 per level) for PG and G in plasma, respectively.

In order to monitor the validity of the measurement, acceptance criteria as set forth in the FDA Guidance for Bioanalytical Method Validation [29] were used. Nicotine and 10 metabolites, namely cotinine, 3-OH-cotinine, nicotine glucuronide, cotinine glucuronide, 3-OH-cotinine glucuronide, 4-OH-4-(3-pyridyl)-butanoic acid, nornicotine, norcotinine, nicotine N-oxide, and cotinine N-oxide were determined by means of solid phase extraction and subsequent LC-MS/MS analysis in urine according to Piller et al. [30] in order to calculate the total nicotine equivalents (TNE).

2.4. Data Evaluation

PG and G values determined in urine or plasma with levels below LLOQ were reported as LLOQ/2. Urinary analyte concentrations (µg/mL) were multiplied with the respective 12 h urine volume to receive the amounts of PG and G excreted within 12 h. The appropriate 12 h urine pools were summed up to obtain the amount of the analytes excreted in 24 h. Evening of the first day until evening of the next day were defined as a 24 h interval (U0 + U1/2, U3 + U4/5, U6 + U7/8). Urinary PG and G excretions were expressed in mg per 24 h (mg/24 h). Means, standard deviations (SD), and medians were calculated, where appropriate. Normal distribution was tested using Shapiro-Wilk and D'Agostino-K squared test. As PG and G concentrations were not normally distributed for the user groups investigated, the non-parametric Mann–Whitney U test (comparison of two groups) and Kruskal–Wallis-ANOVA (comparison of multiple groups) was used to investigate statistical significance between the different nicotine product user and non-user groups. Significance level was set to $\alpha = 0.01$. Correlations were evaluated with the non-parametric Spearman rank correlation analysis. Statistical analysis was conducted in OriginPro 2020b (Version 9.7.5.184, OriginLab Corporation, Northampton, MA, USA).

3. Results and Discussion

A robust method for the determination of PG and G in urine and plasma for human biomonitoring was established and fully validated according to FDA guidelines [29]. For PG, method accuracy rates were between 97.0–101.2% throughout the calibration range. Intra- and inter-day precisions were found to be <10% (CV) in urine (CV < 20% for levels < 3x LLOQ) and <8% (CV) in plasma (CV < 11% for levels < 3x LLOQ). In case of G, method accuracy rates were between 92.0–106.4% throughout the calibration range. Intra- and inter-day precisions were found to be <12% (CV) in urine and <15% (CV) in plasma. Carry-over was monitored by wash injections, whereby no contaminations above LLOQ were identified in this study. Quadratic calibration ranged from 0.1–150 µg/mL for PG and G in urine and plasma, respectively. For additional information with regard to method validation parameters see Supplementary Materials, Table S1.

PG and G were determined in 420 plasma and 360 urine samples of a clinical study. PG and G could be quantified above the LLOQ (0.1 µg/mL) in 360 (100%) and 342 (95%) of the urine and in 217 (52%) and 420 (100%) of the plasma samples, respectively. Representative chromatograms of PG from low and high concentrated urine and plasma samples are shown in Figure 2 (for G, see Figure S1 in Supplementary Materials).

Data within each group were comparable between different days, with exception for the PG level in the first plasma sample, when the subjects were admitted to the clinic (B0, day –1, 5 p.m.). Hence, data from day 3, i.e., the longest time period under confinement and thus under controlled conditions are most predictive of the product-use specific uptake, whereas results from day –1 are assumed to reflect the exposure under real-life conditions caused by the various sources of PG and G (for additional information with regard to day 1 and 2 see Supplementary Materials, Tables S2–S5). Consequently, slightly higher PG concentration in the B0 plasma sample in users of CC, HTP, NRT, OT, and NU compared with levels observed at the following sampling time points for these groups indicate that food and/or the use of daily care products might be a more important source for PG exposure than the use of CC, HTP, NRT and OT products (Figure 3).

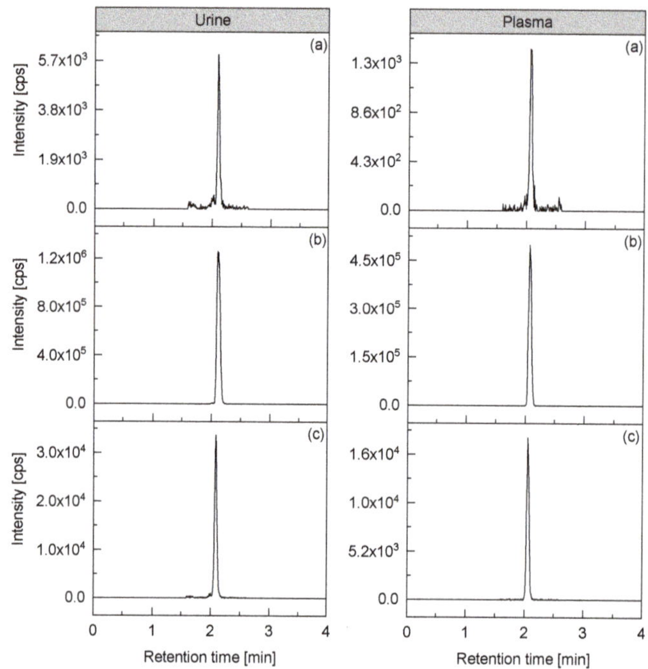

Figure 2. Representative chromatogram of 1,2-propylene glycol (PG) (MRM 285→163) of (**a**) low concentrated urine (0.23 µg/mL PG, user of oral tobacco) and plasma (0.13 µg/mL PG, user of nicotine replacement therapy) samples, (**b**) high concentrated urine (69.9 µg/mL PG, user of e-cigarettes) and plasma (31.9 µg/mL PG, user of e-cigarettes) samples, and (**c**) 1,2-propylene glycol-d$_6$ (MRM 291→169, 2 µg/mL).

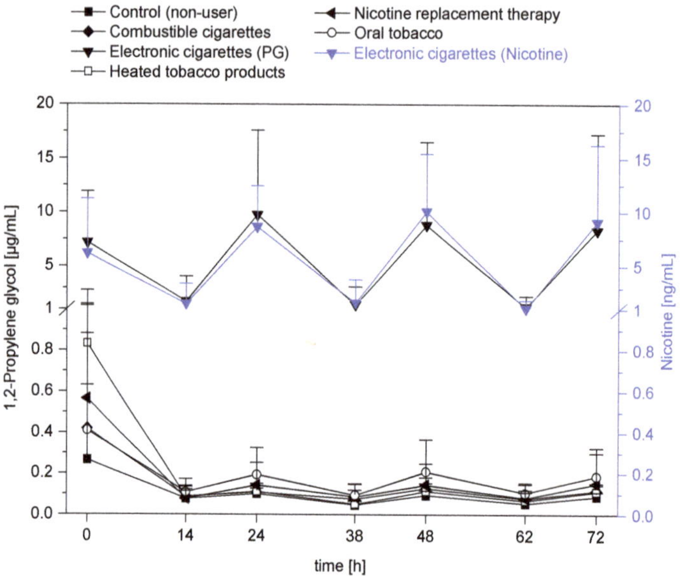

Figure 3. 1,2-Propylene glycol (PG) (µg/mL) as a function of time from day -1 to day 3 in plasma for the different nicotine product user groups and the control. Nicotine (ng/mL) in plasma is shown only for user of electronic cigarettes. Shown is the mean + standard deviation.

Data from analysis of G in plasma samples, with values ranging from 16.5 ± 4.0 µg/mL (mean ± SD) in OT to 21.5 ± 4.6 µg/mL in HTP, showed no significant differences ($p > 0.01$, Kruskal–Wallis-ANOVA) between the different groups investigated (Figure S2a, Table S2 in Supplementary Material) for both time points considered. In comparison, G level in urine were somewhat contradictory, as Kruskal–Wallis-ANOVA showed significant group separation ($p < 0.01$). However, closer examination revealed that the significant results (Mann–Whitney U test, $p < 0.01$) are due to approximately 1.5-fold lower G levels in EC (2.8 ± 0.6 mg/24 h) compared to HTP (4.1 ± 1.1 mg/24 h) users and users of NRT (2.5 ± 1.2 mg/24 h) compared to HTP (4.1 ± 1.1 mg/24 h), OT (4.1 ± 1.3 mg/24 h), and NU (3.8 ± 1.2 mg/24 h), respectively (Figure S2b, Table S2 in Supplementary Materials). In our opinion, the G levels are within the expected background exposure resulting from the daily uptake from food and consumer products [17,23–26] as well as G formed endogenously from proteins, pyruvate, glucose, triacylglycerols, and other glycerolipid metabolic pathways and excreted in urine [31–33]. The findings are in agreement with Landmesser et al. and Nelson et al., showing that oral intake of G below 0.05 g/kg body weight did not result in increased urinary glycerol excretion [28,34]. Data demonstrate that G is not elevated in users of EC, neither in plasma nor urine, and therefore cannot differentiate EC specific uptake of G from other nicotine products.

In contrast, PG levels in plasma showed significant differences ($p < 0.01$, Kruskal–Wallis-ANOVA) between the six groups investigated driven by the elevated levels in users of EC (Table 2, Figure 4a). PG levels in NU (control) ranged from 0.05 ± 0.02 µg/mL (7 a.m., day 3) to 0.09 ± 0.02 µg/mL (5 p.m., day 3) and were comparable to plasma PG levels observed in users of CC, HTP, NRT, and OT (Table 2). Results of the Kruskal–Wallis-ANOVA showed no significant differences between these groups ($p > 0.01$). These results indicate that there is only marginal exposure to PG when using CC, HTP, OT and NRT compared to the background exposure. In contrast, plasma PG levels in EC users were found to be 94-fold elevated in the evening (5 p.m.) and 24-fold elevated in the morning (7 a.m.) compared to NU and up to 71-fold at 5 p.m. and up to 18-fold at 7 a.m. compared to other nicotine product user groups (Table 2, Figure 4a). The observed increase of the PG levels in EC user was significant for both collection time points ($p < 0.01$, Mann–Whitney U test). No significant differences were found between the other groups investigated. Moreover, plasma PG increased significantly (Mann–Whitney U test, $p < 0.01$) throughout the day in EC users with average values of 1.25 µg/mL in the morning and 8.37 µg/mL in the evening (Figure 3). With the exception of the NU group, in which a significant increase of plasma PG from 0.05 µg/mL in the morning to 0.09 µg/mL in the evening was observed, none of the other groups showed an increase in PG levels over day.

Table 2. Descriptive statistics of 1,2-propylene glycol (PG) (µg/mL) in plasma on day 3 in the different nicotine user groups.

User Group	Sampling	N Total	Mean	SD	Median	Min	Max
Control (non-user, NU)	7 a.m.	10	0.05	0.02	0.04	0.03	0.10
	5 p.m.	10	0.09 ***	0.02	0.09	0.05	0.12
Combustible cigarettes (CC)	7 a.m.	10	0.08	0.04	0.06	0.04	0.17
	5 p.m.	10	0.12	0.03	0.12	0.08	0.19
Electronic Cigarettes (EC)	7 a.m.	10	1.25	1.01	0.78	0.25	2.95
	5 p.m.	10	8.37 ***	8.88	5.47	2.38	31.90
Heated Tobacco Products (HTP)	7 a.m.	10	0.07	0.04	0.05	0.03	0.15
	5 p.m.	10	0.11	0.04	0.11	0.05	0.17
Nicotine Replacement Therapy (NRT)	7 a.m.	10	0.08	0.07	0.06	0.02	0.22
	5 p.m.	10	0.15	0.17	0.08	0.04	0.59
Oral Tobacco (OT)	7 a.m.	10	0.11	0.04	0.10	0.06	0.20
	5 p.m.	10	0.19	0.12	0.19	0.07	0.43

SD: standard deviation, Min: minimum, Max: maximum. *** Statistically significant difference of PG between 7 a.m. and 5 p.m. (Mann–Whitney U test, $p < 0.01$).

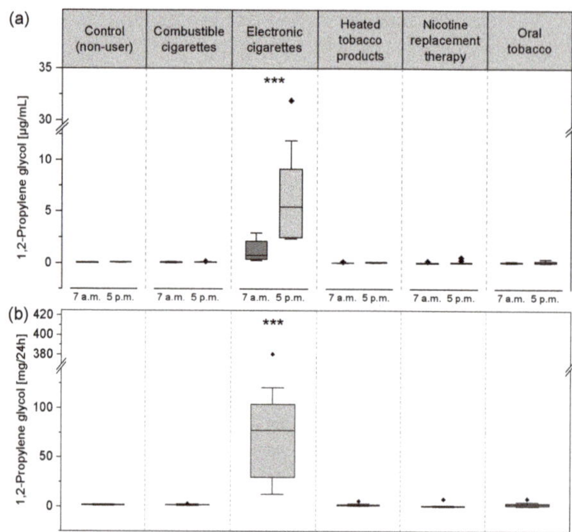

Figure 4. Box-and-whisker plots of 1,2-propylene glycol (PG) on day 3 in (**a**) plasma (µg/mL) at 7 a.m. and 5 p.m. and (**b**) urine (mg/24 h) of different nicotine product user groups and the control. Box-and-whisker plots represent medians (horizontal lines) with 25% and 75% percentiles (boxes), 1.5 × IQR (whiskers), and outliers (diamonds). *** $p < 0.01$ (Mann–Whitney U test, comparison of EC with each other group).

Similar results were observed for PG in urine showing significantly elevated levels in EC users as compared to all other four nicotine user groups and the control group ($p < 0.01$, Kruskal–Wallis-ANOVA). PG in EC users showed levels of 95.4 ± 107.1 mg/24 h, 62 times higher than in NU (control) and between 29 to 54 times increased compared to the other nicotine user groups (Table 3, Figure 4b). Differences between EC and the other groups were statistically significant ($p = 1.1 \times 10^{-5}$, Mann–Whitney U test). The PG amount detected in the control group (NU) was in the range of 1.5 ± 0.4 mg/24 h (mean ± SD, Table 3, Figure 4b). No significant differences ($p < 0.01$, Mann–Whitney U test) were observed between the NU and users of CC (1.8 ± 0.5 mg/24 h), HTP (2.1 ± 1.5 mg/24 h), OT (3.3 ± 2.3 mg/24 h), and NRT (1.7 ± 2.3 mg/24 h). Kruskal–Wallis-ANOVA of these groups, excluding EC users, confirmed the results and showed no significant differences ($p > 0.01$).

Table 3. Descriptive statistics of 1,2-propylene glycol (PG) (mg/24 h) in urine on day 3 in the different nicotine user groups.

User Group	N Total	Mean	SD	Median	Min	Max
Control (non-user, NU)	10	1.5	0.4	1.5	1.0	2.1
Combustible cigarettes (CC)	10	1.8	0.5	1.7	1.0	2.7
Electronic cigarettes (EC)	10	95.4 ***	107.1	77.6	12.7	380.7
Heated tobacco products (HTP)	10	2.1	1.5	1.6	0.9	5.9
Nicotine replacement therapy (NRT)	10	1.7	2.3	1.1	0.4	8.1
Oral tobacco (OT)	10	3.3	2.3	3.0	1.1	8.7

SD: standard deviation, Min: minimum, Max: maximum; ***: Statistically significant from all other groups ($p < 0.01$).

These results clearly demonstrate a separation between EC users and other nicotine product user groups in terms of PG levels in plasma and urine. Plasma PG concentrations and urinary PG excretion on day -1 (B0 and U0 samples) confirm these results, even though with higher variability, especially for users of HTP, NRT, OT (Figure S3 and Table S6 in Supplementary Materials). Also, under uncontrolled real-life conditions, PG levels in users of EC were significantly elevated compared with the other nicotine product user groups and the control ($p < 0.01$, Mann–Whitney U test). PG levels in EC users were found to be increased 9- to 27-fold in plasma and 12- to 32-fold in urine samples, clearly demonstrating a group separation of EC users and other nicotine product user groups based on PG levels under uncontrolled conditions.

Although PG levels were only highly increased in urine and plasma samples of EC users, a general trend can also be observed for the other groups investigated: NU < CC, HTP, and NRT < OT << EC. The low PG levels detected in the control group (NU) in both urine and plasma samples, reflect the background concentrations, most probably from daily use of consumer products [17,23–26]. In relation to the excreted volume, PG levels in urine correspond to 672 ± 209 ng/mL, not significantly different (Mann–Whitney U test, $p > 0.01$) from Wurita et al. that observed PG levels of 5450 ± 9290 ng/mL (mean $\pm$ SD, range: 491–41600 ng/mL) in urine of 23 healthy subjects [35]. In addition, if PG was normalized to creatinine (determined as part of the project), the PG level in the control group (NU) was 1.6 ± 0.6 mmol/mol creatinine, which is comparable with the value of 2.3 ± 1.4 mmol/mol creatinine observed by Laitinen et al. [36]. A statement regarding plasma levels is not possible, as data on the general background exposure to PG in plasma are lacking in the literature. Slightly higher PG concentrations detected in users of CC, HTP, and NRT could be attributed to the addition of PG as humectant to conventional cigarettes [37–40], HTP [41,42], and nicotine sprays (e.g., Nicorette® Mint Spray) used for NRT. In comparison, even higher PG levels were observed in users of OT such as snus products, to which PG is also added as a humectant [43]. As mentioned before, the EC users show by far the highest PG values, which can be attributed to PG intake from EC use [44,45]. High variations in the group of EC users in urine and plasma can be ascribed to various factors affecting the PG uptake, such as varying PG contents in the used e-liquids, e-cigarette device characteristics such as model, wattage, and temperature, as well as the use behavior of the individual subject including puff number, puff volume, puff duration and depth of inhalation [46–55]. In fact, measured PG level highly depend on the PG content of the liquid consumed. However, as the plasma and urine PG levels were >9-fold elevated in EC users, we assume that the critical level to differentiate between EC user and other consumer groups would be a PG content <10%. Theoretical mathematical assessment with 5 times lower PG level in EC users still holds statistical significance substantiating the limit 10% PG content. We were able to demonstrate that the EC specific PG uptake is better reflected by its plasma concentrations measured at different time points, which showed a 6.7-fold increase throughout the day. Lower plasma PG levels in the morning can be attributed to the prohibition of EC use from midnight to 9 a.m. applied in the clinical study, leading to a partial washout within this time frame (Figure 3). The increase in plasma PG concentrations during the day was already observed by Landmesser et al. and can be unequivocally attributed to EC use [28]. The correlation of excreted PG with the actual amount of PG inhaled by the individual user would be of utmost interest, as it would contribute to the comprehensive assessment of PG intake after EC use [56,57]. Therefore, the PG intake was calculated as an estimate according to Equation (1), where I_{PG} is the intake of PG (g/d) from EC use, LPD is the amount of e-liquid (g) consumed per day, MSp is the mouth spill, and R is the respiratory retention. As a first approximation, a PG content of 50%, a mouth spill of 30% [58], and a respiratory retention of 92% [59] was assumed.

$$I_{PG} = LPD \times 0.5 \times (1 - MSP) \times R \tag{1}$$

The urinary PG excretion (mg/24 h) showed a significant ($p < 0.01$) correlation with the calculated PG intake with a Spearman's rank correlation coefficient of $r = 0.976$ (Figure 5a). Additionally, the association of vaping-related PG uptake was indirectly addressed by correlating the PG biomarker levels in EC users with specific biomarkers that provide a measure of product consumption. Urinary PG measured on day 3 was compared with TNE, the molar sum of nicotine and 10 metabolites [60]. As expected, the urinary PG concentrations showed a correlation with the excreted amount of TNE with a Spearman's rank correlation coefficient of $r = 0.66$ ($p = 0.038$, Figure 5b) indicating that more PG is absorbed and excreted upon higher uptake of EC aerosol. Plasma PG levels correlate only moderately with nicotine plasma levels on day 3 for both collection time points (7 a.m. and 5 p.m.), with a Spearman's rank correlation coefficient of r = 0.37 for each (Figure 5c). Although the correlation of plasma PG with nicotine is not significant, a clear relationship can be observed when considering both collection time points each (Figure 5c) and the plasma time course (Figure 3). This correlation could be explained by comparable plasma half-lives of nicotine (approx. 2 h after inhalation [61–63]) and PG (2.3 ± 0.7 h after intravenous infusion [64] and 3.8 ± 0.8 h/4.1 ± 0.7 h after oral administration [65], and approx. 2 h after vaping [28]), confirming previous findings [28].

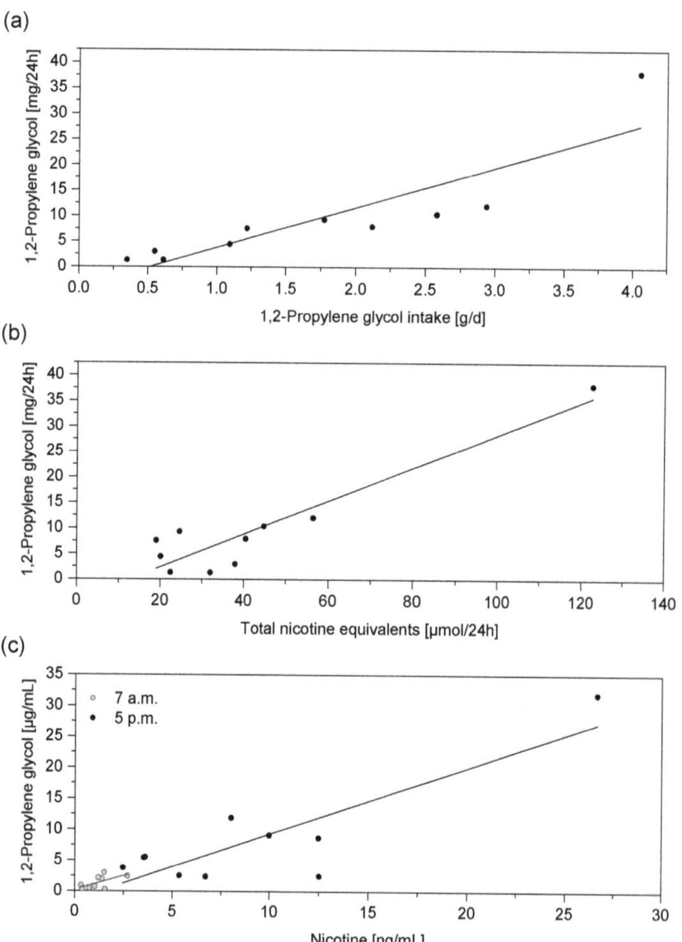

Figure 5. (**a**) Correlation between 1,2-propylene glycol (PG) in urine (mg/24 h) and the estimated PG intake (g/d) of EC users on day 3 (Spearman's correlation: r = 0.976 ($p = 1.4 \times 10^{-6}$), (**b**) correlation between PG in urine (mg/24 h) and urinary nicotine equivalents (µmol/24 h) of EC users (Spearman's correlation r = 0.661, $p = 0.0376$), (**c**) correlation of PG (µg/mL) and nicotine (ng/mL) in plasma of EC users on day 3 at 7 a.m. with a Spearman's correlation r = 0.370 ($p = 0.293$), and at 5 p.m. with a Spearman's correlation r = 0.370 ($p = 0.293$).

The daily PG intake of the EC users was in average 1.7 g PG/d (range: 0.3–4.1 g/d, estimated with Equation (1)). Mean plasma PG concentrations on the 3 study days at 5 p.m. were in the range of 8.4–9.8 µg/mL. These data are roughly comparable with a study from Speth et al., in which a PG dose of 5.1 g was administered intravenously (IV) to 3 subjects over a time period of 4 h [64]. PG plasma concentrations were reported to be in the range of 48–60 µg/mL. While the dose in the IV study was on average three times higher compared to our study, the plasma levels were about 6-fold increased. This apparent discrepancy is best explained by the time period of dosing: 4 h in the IV study versus ~8 h in the present study. Interestingly, Speth et al. [64] reported a saturable PG clearance over the applied dose range of 5–21 g/d. Whether this phenomenon is also relevant of the vaping-related doses remains to be investigated. Finally, it is noteworthy that these authors found no evidence for lactate acidosis, hemolysis or increase in osmolality in the studied dose range [64].

An obstacle for using PG (in plasma or urine) as a biomarker of exposure might be the inter-individual variability in pharmacokinetic parameters such as half-life, volume of distribution and achievable plasma levels as reported for humans in the literature [64,65]. However, the strong dose-response relationships observed for urinary excretion of PG in vapers (Figure 5a,b) indicate that this is unlikely to be a major issue.

Although the emergence of EC in the last decade led to the need to identify a biomarker specific for EC use [5], most studies have focused on biomarkers of tobacco-smoke exposure, which are not specific to EC use. Moreover, studies often only comprise the comparison of EC to NU and/or CC alone [5,66–71]. Only a few studies include other nicotine product user groups [72,73]. To the best of our knowledge, this is the first study systematically assessing the PG and G concentration in plasma and urine samples in users of EC and four additional nicotine product user groups, namely users of CC, HTP, NRT, and OT. G levels did not differ between the different groups investigated, neither in urine nor in plasma. The reason for this is the high and variable background level of G in plasma and urine caused by a common exposure to G (food, consumer products) and the endogenous formation of G in the lipid metabolism. Therefore, G is not suitable as a specific biomarker to identify EC use. In contrast, levels of PG were significantly elevated in users of EC compared to the control group and all nicotine user groups investigated, despite the small sample size (10 subject per group), which represents a limitation of the current study. Larger studies under field conditions are required to support the suitability of PG in plasma or urine as specific biomarker for the use of ECs. Another limitation is a minimum amount of presumable ≥10% PG in the e-liquid, to detect a difference between users of EC and other nicotine user groups. However, further investigations are needed in order to verify this cut-off.

4. Conclusions

In conclusion, the current study clearly demonstrates a significant distinction between users of EC and users of CC, HTP, NRT, OT as well as NU based on the PG level in urine and plasma. The observed dose-response relationship between urinary and plasma PG and intensity of vaping (daily consumption and nicotine uptake) emphasizes the suitability of PG as a potential biomarker of EC use. Due to the restricted sample size of the current study, we recommend verifying these results under field conditions. Consequently, we propose the use of PG in urine and/or plasma in order to monitor EC use compliance in exposure assessments under real-life conditions (field and epidemiological studies). Moreover, a combination of several biomarkers may lead to a more comprehensive differentiation among several user groups which would provide a better understanding of exposure and related health effects.

Supplementary Materials: The following are available online at https://www.mdpi.com/article/10.3390/separations8100180/s1, Table S1. Method validation parameters for the determination of propylene glycol and glycerol in urine and plasma, Figure S1. Representative chromatogram of glycerol (G) (MRM 405→283) of (a) low concentrated urine (0.21 µg/mL G, user of oral tobacco) and plasma (9.9 µg/mL G, non-user) samples, (b) high concentrated urine (9.7 µg/mL G, user of e-cigarettes) and plasma (24.4 µg/mL PG, user of e-cigarettes) samples, and (c) glycerol-d5 (MRM 410→288, 2 µg/mL), Figure S2. Box-and-whisker plots of glycerol (G) on day 3 in (a) plasma (µg/mL) at 7 a.m. and 5 p.m. and (b) urine (mg/24 h) between different nicotine product user groups and the control. Box-and-whisker plots represent medians (horizontal lines) with 25% and 75% percentiles (boxes), 1.5xIQR (whiskers), and outliers (diamonds), Table S2. Descriptive statistics of glycerol (µg/mL) in plasma of study samples for different nicotine user groups, Table S3. Descriptive statistics of glycerol (mg/24 h) in urine of study samples for different nicotine user groups, Table S4. Descriptive statistics of 1,2-propylene glycol (µg/mL) in plasma of study samples for different nicotine user groups, Table S5. Descriptive statistics of PG (mg/24 h) in urine of study samples for different nicotine user groups, Figure S3. Box-and-whisker plot of 1,2-propylene glycol (PG) on day −1 in (a) plasma (µg/mL) at 5 p.m. (B0) and (b) urine (mg/12 h) (U0) between different nicotine product user groups and the control. Box-and-whisker plots represent medians (horizontal lines) with 25% and 75% percentiles (boxes), 1.5xIQR (whiskers), and outliers (diamonds). *** $p < 0.01$ (comparison of EC vs all other groups), Table S6. Descriptive statistics of PG in urine (mg/12 h) and plasma (µg/mL) of study samples for different nicotine user groups at day -1.

Author Contributions: Conceptualization, M.S., G.S. and N.P.; formal analysis, T.B.; writing— original draft preparation, T.B.; writing—review and editing, N.P., M.S. and G.S.; supervision, M.S.; project administration, M.S.; funding acquisition, M.S. and N.P. All authors have read and agreed to the published version of the manuscript.

Funding: This study was funded with a grant from the Foundation for a Smoke-Free World, a US nonprofit 501(c)(3) private foundation with a mission to end smoking in this generation. The Foundation accepts charitable gifts from PMI Global Services Inc. (PMI); under the Foundation's Bylaws and Pledge Agreement with PMI, the Foundation is independent from PMI and the tobacco industry. The contents, selection, and presentation of facts, as well as any opinions expressed herein are the sole responsibility of the authors and under no circumstances shall be regarded as reflecting the positions of the Foundation for a Smoke-Free World, Inc.

Institutional Review Board Statement: The study was conducted according to the guidelines of the Declaration of Helsinki, and was approved by the Ethics Committee of the Medical Association Hamburg (reference number: PV7084, date of approval: 10 September 2019).

Informed Consent Statement: Informed consent was obtained from all subjects involved in the study.

Data Availability Statement: The data presented here are available upon request from the corresponding author.

Acknowledgments: The authors thank CTC North GmbH (Hamburg, Germany) for conducting the clinical study, and our colleagues from ABF for their assistance in organizing and managing the clinical study, and performing the sample analysis.

Conflicts of Interest: The authors declare no conflict of interest. The funders had no role in the design of the study; in the collection, analyses, or interpretation of data; in the writing of the manuscript, or in the decision to publish the results.

References

1. Institute of Medicine (IOM). *Clearing the Smoke: Assessing the Science Base for Tobacco Harm Reduction*; National Academy Press: Washington, DC, USA, 2001; pp. 1–500.
2. US Department of Health and Human Services. *How Tobacco Smoke Causes Disease: The Biology and Behavioral Basis for Smoking-Attributable Disease—A Report of the Surgeon General*; National Library of Medicine Cataloging in Publication: Rockville, MD, USA, 2010; pp. 1–792.
3. Djordjevic, M.V.; Doran, K.A. Nicotine content and delivery across tobacco products. *Handb. Exp. Pharmacol.* **2009**, 61–82. [CrossRef]
4. Chang, C.M.; Edwards, S.H.; Arab, A.; Del Valle-Pinero, A.Y.; Yang, L.; Hatsukami, D.K. Biomarkers of Tobacco Exposure: Summary of an FDA-Sponsored Public Workshop. *Cancer Epidemiol. Biomark. Prev. Publ. Am. Assoc. Cancer Res. Cosponsored Am. Soc. Prev. Oncol.* **2016**. [CrossRef]
5. Schick, S.F.; Blount, B.C.; Jacob, P., 3rd; Saliba, N.A.; Bernert, J.T.; El Hellani, A.; Jatlow, P.; Pappas, R.S.; Wang, L.; Foulds, J.; et al. Biomarkers of Exposure to New and Emerging Tobacco and Nicotine Delivery Products. *Am. J. Physiology Lung Cell. Mol. Physiol.* **2017**, *313*, L425–L452. [CrossRef]
6. Alwis, K.U.; Blount, B.C.; Britt, A.S.; Patel, D.; Ashley, D.L. Simultaneous analysis of 28 urinary VOC metabolites using ultra high performance liquid chromatography coupled with electrospray ionization tandem mass spectrometry (UPLC-ESI/MSMS). *Anal. Chim. Acta* **2012**, *750*, 152–160. [CrossRef]

7. Pluym, N.; Gilch, G.; Scherer, G.; Scherer, M. Analysis of 18 urinary mercapturic acids by two high-throughput multiplex-LC-MS/MS methods. *Anal. Bioanal. Chem.* **2015**, *407*, 5463–5476. [CrossRef]

8. Xia, B.; Blount, B.C.; Guillot, T.; Brosius, C.; Li, Y.; Van Bemmel, D.M.; Kimmel, H.L.; Chang, C.M.; Borek, N.; Edwards, K.C.; et al. Tobacco-Specific Nitrosamines (NNAL, NNN, NAT, and NAB) Exposures in the US Population Assessment of Tobacco and Health (PATH) Study Wave 1 (2013-2014). *Nicotine Tob. Res. Off. J. Soc. Res. Nicotine Tob.* **2021**, *23*, 573–583. [CrossRef]

9. Riedel, K.; Scherer, G.; Engl, J.; Hagedorn, H.W.; Tricker, A.R. Determination of three carcinogenic aromatic amines in urine of smokers and nonsmokers. *J. Anal. Toxicol.* **2006**, *30*, 187–195. [CrossRef]

10. McNeill, A.; Brose, L.S.; Calder, R.; Hitchman, S.C.; Hajek, P.; McRobbie, H. *E-cigarettes: An Evidence Update. A Report Commissioned by Public Health England*; Public Health England: London, UK, 2015; pp. 1–113.

11. McNeill, A.; Brose, L.; Calder, R.; Simonavicius, E.; Robson, D. *Vaping in England: An Evidence Update including Vaping for Smoking Cessation, February 2021*; Public Health England: London, UK, 2021; pp. 1–247.

12. Hatsukami, D.K.; Carroll, D.M. Tobacco harm reduction: Past history, current controversies and a proposed approach for the future. *Prev. Med.* **2020**, *140*, 106099. [CrossRef]

13. Adkison, S.E.; O'Connor, R.J.; Bansal-Travers, M.; Hyland, A.; Borland, R.; Yong, H.H.; Cummings, K.M.; McNeill, A.; Thrasher, J.F.; Hammond, D.; et al. Electronic nicotine delivery systems: International tobacco control four-country survey. *Am. J. Prev. Med.* **2013**, *44*, 207–215. [CrossRef]

14. Mayer, M.; Reyes-Guzman, C.; Grana, R.; Choi, K.; Freedman, N.D. Demographic Characteristics, Cigarette Smoking, and e-Cigarette Use Among US Adults. *JAMA Netw. Open* **2020**, *3*, e2020694. [CrossRef]

15. Kapan, A.; Stefanac, S.; Sandner, I.; Haider, S.; Grabovac, I.; Dorner, T.E. Use of Electronic Cigarettes in European Populations: A Narrative Review. *Int. J. Environ. Res. Public Health* **2020**, *17*, 1917. [CrossRef]

16. Prakash, S.; Hatcher, C.; Shiffman, S. Prevalence of ENDS and JUUL Use, by Smoking Status, in National Samples of Young Adults and Older Adults in the U.S. *Am. J. Health Behav.* **2021**, *45*, 402–418. [CrossRef]

17. Bonner, E.; Chang, Y.; Christie, E.; Colvin, V.; Cunningham, B.; Elson, D.; Ghetu, C.; Huizenga, J.; Hutton, S.J.; Kolluri, S.K.; et al. The chemistry and toxicology of vaping. *Pharmacol. Ther.* **2021**, *225*, 107837. [CrossRef]

18. Dai, J.; Kim, K.-H.; Szulejko, J.E.; Jo, S.-H.; Kwon, K.; Choi, D.W. Quantification of nicotine and major solvents in retail electronic cigarette fluids and vaped aerosols. *Microchem. J.* **2018**, *140*, 262–268. [CrossRef]

19. Crenshaw, M.D.; Tefft, M.E.; Buehler, S.S.; Brinkman, M.C.; Clark, P.I.; Gordon, S.M. Determination of Nicotine, Glycerol, Propylene Glycol and Water in Electronic Cigarette Fluids Using Quantitative 1 H NMR. *Magn. Reson. Chem. MRC* **2016**, *54*, 901–904. [CrossRef]

20. Ruddick, J.A. Toxicology, metabolism, and biochemistry of 1, 2-propanediol. *Toxicol. App. Pharmacol.* **1972**, *21*, 102–111. [CrossRef]

21. Wilson, K.C.; Reardon, C.; Theodore, A.C.; Farber, H.W. Propylene glycol toxicity: A severe iatrogenic illness in ICU patients receiving IV benzodiazepines: A case series and prospective, observational pilot study. *Chest* **2005**, *128*, 1674–1681. [CrossRef]

22. Koehler, K.; Braun, H.; de Marees, M.; Geyer, H.; Thevis, M.; Mester, J.; Schaenzer, W. Urinary excretion of exogenous glycerol administration at rest. *Drug Test. Anal.* **2011**, *3*, 877–882. [CrossRef]

23. McMartin, K. Propylene Glycol. In *Encyclopedia of Toxicology*, 3rd ed.; Wexler, P., Ed.; Academic Press: Oxford, UK, 2014; pp. 1113–1116.

24. Hutzler, C.; Paschke, M.; Kruschinski, S.; Henkler, F.; Hahn, J.; Luch, A. Chemical hazards present in liquids and vapors of electronic cigarettes. *Arch. Toxicol.* **2014**, *88*, 1295–1308. [CrossRef]

25. National Toxicology Program. NTP-CERHR Monograph on the Potential Human Reproductive and Developmental Effects of Propylene Glycol (PG). *NTP CERHR Mon.* **2004**, *12*, 1–III6.

26. Wernke, M.J. Glycerol. In *Encyclopedia of Toxicology*, 3rd ed.; Wexler, P., Ed.; Academic Press: Oxford, UK, 2014; pp. 754–756.

27. Sibul, F.; Burkhardt, T.; Kachhadia, A.; Pilz, F.; Scherer, G.; Scherer, M.; Pluym, N. Identification of biomarkers specific to five different nicotine product user groups: Study protocol of a controlled clinical trial. *Contemp. Clin. Trials Commun.* **2021**, *22*, 100794. [CrossRef]

28. Landmesser, A.; Scherer, M.; Pluym, N.; Sarkar, M.; Edmiston, J.; Niessner, R.; Scherer, G. Biomarkers of Exposure Specific to E-vapor Products Based on Stable-Isotope Labeled Ingredients. *Nicotine Tob. Res. Off. J. Soc. Res. Nicotine Tob.* **2019**, *21*, 314–322. [CrossRef]

29. Food and Drug Administration (FDA). Bioanalytical Method Validation—Guidance for Industry. 2018. Available online: https://www.fda.gov/downloads/Drugs/GuidanceComplianceRegulatoryInformation/Guidances/UCM070107.pdf (accessed on 12 August 2021).

30. Piller, M.; Gilch, G.; Scherer, G.; Scherer, M. Simple, fast and sensitive LC-MS/MS analysis for the simultaneous quantification of nicotine and 10 of its major metabolites. *J. Chromatogr. B: Anal. Technol. Biomed. Life Sci.* **2014**, *951–952*, 7–15. [CrossRef]

31. Jansson, P.A.; Larsson, A.; Smith, U.; Lönnroth, P. Glycerol production in subcutaneous adipose tissue in lean and obese humans. *J. Clin. Investig.* **1992**, *89*, 1610–1617. [CrossRef]

32. Rotondo, F.; Ho-Palma, A.C.; Remesar, X.; Fernández-López, J.A.; Romero, M.d.M.; Alemany, M. Glycerol is synthesized and secreted by adipocytes to dispose of excess glucose, via glycerogenesis and increased acyl-glycerol turnover. *Sci. Rep.* **2017**, *7*, 8983. [CrossRef]

33. Brisson, D.; Vohl, M.-C.; St-Pierre, J.; Hudson, T.J.; Gaudet, D. Glycerol: A neglected variable in metabolic processes? *BioEssays* **2001**, *23*, 534–542. [CrossRef]

34. Nelson, J.L.; Harmon, M.E.; Robergs, R.A. Identifying plasma glycerol concentration associated with urinary glycerol excretion in trained humans. *J. Anal. Toxicol.* **2011**, *35*, 617–623. [CrossRef]

35. Wurita, A.; Suzuki, O.; Hasegawa, K.; Gonmori, K.; Minakata, K.; Yamagishi, I.; Nozawa, H.; Watanabe, K. Presence of appreciable amounts of ethylene glycol, propylene glycol, and diethylene glycol in human urine of healthy subjects. *Forensic Toxicol.* **2014**, *32*, 39–44. [CrossRef]

36. Laitinen, J.; Liesivuori, J.; Savolainen, H. Exposure to glycols and their renal effects in motor servicing workers. *Occup. Med.* **1995**, *45*, 259–262. [CrossRef]

37. Jansen, E.; Ramlal, R.; Cremers, H.; Talhout, R. Stability and Concentrations of Humectants in Tobacco. *J. Anal. Bioanal. Tech.* **2017**, *8*, 380. [CrossRef]

38. Pennings, J.L.A.; Cremers, J.; Becker, M.J.A.; Klerx, W.N.M.; Talhout, R. Aldehyde and VOC yields in commercial cigarette mainstream smoke are mutually related and depend on the sugar and humectant content in tobacco. *Nicotine Tob. Res. Off. J. Soc. Res. Nicotine Tob.* **2019**, *10*, 748–1756. [CrossRef]

39. Heck, J.D.; Gaworski, C.L.; Rajendran, N.; Morrissey, R.L. Toxicologic evaluation of humectants added to cigarette tobacco: 13-week smoke inhalation study of glycerin and propylene glycol in Fischer 344 rats. *Inhal. Toxicol.* **2002**, *14*, 1135–1152. [CrossRef]

40. Gaworski, C.L.; Oldham, M.J.; Coggins, C.R.E. Toxicological considerations on the use of propylene glycol as a humectant in cigarettes. *Toxicology* **2010**, *269*, 54–66. [CrossRef]

41. Stepanov, I.; Woodward, A. Heated tobacco products: Things we do and do not know. *Tob. Control* **2018**, *27*, s7–s8. [CrossRef] [PubMed]

42. Uchiyama, S.; Noguchi, M.; Takagi, N.; Hayashida, H.; Inaba, Y.; Ogura, H.; Kunugita, N. Simple Determination of Gaseous and Particulate Compounds Generated from Heated Tobacco Products. *Chem. Res. Toxicol.* **2018**, *7*, 585–593. [CrossRef]

43. McAdam, K.G.; Kimpton, H.; Faizi, A.; Porter, A.; Rodu, B. The composition of contemporary American and Swedish smokeless tobacco products. *BMC Chem.* **2019**, *13*, 31. [CrossRef] [PubMed]

44. Eaton, D.L.; Kwan, L.Y.; Stratton, K.; National Academies of Sciences, E. Toxicology of E-Cigarette Constituents. In *Public Health Consequences of E-Cigarettes*; National Academies Press (US): Washington, DC, USA, 2018.

45. Hajek, P.; Etter, J.F.; Benowitz, N.; Eissenberg, T.; McRobbie, H. Electronic cigarettes: Review of use, content, safety, effects on smokers and potential for harm and benefit. *Addiction* **2014**, *109*, 1801–1810. [CrossRef] [PubMed]

46. Gillman, I.G.; Kistler, K.A.; Stewart, E.W.; Paolantonio, A.R. Effect of variable power levels on the yield of total aerosol mass and formation of aldehydes in e-cigarette aerosols. *Regul. Toxicol. Pharmacol.* **2016**, *75*, 58–65. [CrossRef] [PubMed]

47. St Helen, G.; Shahid, M.; Chu, S.; Benowitz, N.L. Impact of e-liquid flavors on e-cigarette vaping behavior. *Drug Alcohol Depend.* **2018**, *189*, 42–48. [CrossRef] [PubMed]

48. Cunningham, A.; Slayford, S.; Vas, C.; Gee, J.; Costigan, S.; Prasad, K. Development, validation and application of a device to measure e-cigarette users' puffing topography. *Sci. Rep.* **2016**, *6*, 35071. [CrossRef]

49. Spindle, T.R.; Talih, S.; Hiler, M.M.; Karaoghlanian, N.; Halquist, M.S.; Breland, A.B.; Shihadeh, A.; Eissenberg, T. Effects of electronic cigarette liquid solvents propylene glycol and vegetable glycerin on user nicotine delivery, heart rate, subjective effects, and puff topography. *Drug Alcohol Depend.* **2018**, *188*, 193–199. [CrossRef] [PubMed]

50. Talih, S.; Balhas, Z.; Eissenberg, T.; Salman, R.; Karaoghlanian, N.; El Hellani, A.; Baalbaki, R.; Saliba, N.; Shihadeh, A. Effects of User Puff Topography, Device Voltage, and Liquid Nicotine Concentration on Electronic Cigarette Nicotine Yield: Measurements and Model Predictions. *Nicotine Tob. Res. Off. J. Soc. Res. Nicotine Tob.* **2014**, *3*, 150–157. [CrossRef] [PubMed]

51. DeVito, E.E.; Krishnan-Sarin, S. E-cigarettes: Impact of E-Liquid Components and Device Characteristics on Nicotine Exposure. *Curr. Neuropharmacol.* **2018**, *16*, 438–459. [CrossRef] [PubMed]

52. El-Hellani, A.; Salman, R.; El-Hage, R.; Talih, S.; Malek, N.; Baalbaki, R.; Karaoghlanian, N.; Nakkash, R.; Shihadeh, A.; Saliba, N.A. Nicotine and Carbonyl Emissions From Popular Electronic Cigarette Products: Correlation to Liquid Composition and Design Characteristics. *Nicotine Tob. Res. Off. J. Soc. Res. Nicotine Tob.* **2016**, *20*, 215–223. [CrossRef]

53. Farsalinos, K.E.; Spyrou, A.; Tsimopoulou, K.; Stefopoulos, C.; Romagna, G.; Voudris, V. Nicotine absorption from electronic cigarette use: Comparison between first and new-generation devices. *Sci. Rep.* **2014**, *4*, 4133. [CrossRef] [PubMed]

54. Li, Y.; Burns, A.E.; Tran, L.N.; Abellar, K.A.; Poindexter, M.; Li, X.; Madl, A.K.; Pinkerton, K.E.; Nguyen, T.B. Impact of e-Liquid Composition, Coil Temperature, and Puff Topography on the Aerosol Chemistry of Electronic Cigarettes. *Chem. Res. Toxicol.* **2021**, *34*, 1640–1654. [CrossRef]

55. Cahours, X.; Prasad, K. A Review of Electronic Cigarette Use Behaviour Studies. *Beiträge Zur Tab. Int. Contrib. Tob. Res.* **2018**, *28*, 81–92. [CrossRef]

56. Rebuli, M.E.; Liu, F.; Urman, R.; Barrington-Trimis, J.L.; Eckel, S.P.; McConnell, R.; Jaspers, I. Compliance in controlled e-cigarette studies. *Nicotine Tob. Res. Off. J. Soc. Res. Nicotine Tob.* **2020**, 17. [CrossRef]

57. Winden, T.J.; Chen, E.S.; Wang, Y.; Sarkar, I.N.; Carter, E.W.; Melton, G.B. Towards the Standardized Documentation of E-Cigarette Use in the Electronic Health Record for Population Health Surveillance and Research. *AMIA Jt Summits Transl. Sci. Proc.* **2015**, *2015*, 199–203.

58. Logue, J.M.; Sleiman, M.; Montesinos, V.N.; Russell, M.L.; Litter, M.I.; Benowitz, N.L.; Gundel, L.A.; Destaillats, H. Emissions from Electronic Cigarettes: Assessing Vapers' Intake of Toxic Compounds, Secondhand Exposures, and the Associated Health Impacts. *Environ. Sci. Technol.* **2017**, *51*, 9271–9279. [CrossRef]

59. St Helen, G.; Havel, C.; Dempsey, D.A.; Jacob, P., 3rd; Benowitz, N.L. Nicotine delivery, retention and pharmacokinetics from various electronic cigarettes. *Addiction* **2016**, *111*, 535–544. [CrossRef]

60. Benowitz, N.L.; St Helen, G.; Nardone, N.; Cox, L.S.; Jacob, P. Urine Metabolites for Estimating Daily Intake of Nicotine from Cigarette Smoking. *Nicotine Tob. Res. Off. J. Soc. Res. Nicotine Tob.* **2020**, *22*, 288–292. [CrossRef] [PubMed]

61. Holloway, A.C. Nicotine. In *Encyclopedia of Toxicology*, 3rd ed.; Wexler, P., Ed.; Academic Press: Oxford, UK, 2014; pp. 514–516.

62. Feyerabend, C.; Ings, R.M.J.; Russell, M.A.H. Nicotine pharmacokinetics and its application to intake from smoking. *Br. J. Clin. Pharmacol.* **1985**, *19*, 239–247. [CrossRef] [PubMed]

63. Benowitz, N.L.; Hukkanen, J.; Jacob, P., 3rd. Nicotine chemistry, metabolism, kinetics and biomarkers. *Handb. Exp. Pharmacol.* **2009**, 29–60. [CrossRef]

64. Speth, P.A.; Vree, T.B.; Neilen, N.F.; de Mulder, P.H.; Newell, D.R.; Gore, M.E.; de Pauw, B.E. Propylene glycol pharmacokinetics and effects after intravenous infusion in humans. *Ther. Drug Monit.* **1987**, *9*, 255–258. [CrossRef]

65. Yu, D.K.; Elmquist, W.F.; Sawchuk, R.J. Pharmacokinetics of Propylene Glycol in Humans During Multiple Dosing Regimens. *J. Pharm. Sci.* **1985**, *74*, 876–879. [CrossRef] [PubMed]

66. Conklin, D.J.; Ogunwale, M.A.; Chen, Y.; Theis, W.S.; Nantz, M.H.; Fu, X.A.; Chen, L.C.; Riggs, D.W.; Lorkiewicz, P.; Bhatnagar, A.; et al. Electronic cigarette-generated aldehydes: The contribution of e-liquid components to their formation and the use of urinary aldehyde metabolites as biomarkers of exposure. *Aerosol Sci. Technol. J. Am. Assoc. Aerosol Res.* **2018**, *52*, 1219–1232. [CrossRef]

67. Sakamaki-Ching, S.; Williams, M.; Hua, M.; Li, J.; Bates, S.M.; Robinson, A.N.; Lyons, T.W.; Goniewicz, M.L.; Talbot, P. Correlation between biomarkers of exposure, effect and potential harm in the urine of electronic cigarette users. *BMJ Open Respir. Res.* **2020**, *7*, e000452. [CrossRef]

68. Goniewicz, M.L.; Smith, D.M.; Edwards, K.C.; Blount, B.C.; Caldwell, K.L.; Feng, J.; Wang, L.; Christensen, C.; Ambrose, B.; Borek, N.; et al. Comparison of Nicotine and Toxicant Exposure in Users of Electronic Cigarettes and Combustible Cigarettes. *JAMA Net. Open* **2018**, *1*, e185937. [CrossRef]

69. Glasser, A.M.; Collins, L.; Pearson, J.L.; Abudayyeh, H.; Niaura, R.S.; Abrams, D.B.; Villanti, A.C. Overview of Electronic Nicotine Delivery Systems: A Systematic Review. *Am. J. Prev. Med.* **2016**, *52*, e33–e66. [CrossRef]

70. Faridoun, A.; Sultan, A.S.; Jabra-Rizk, M.A.; Weikel, D.; Varlotta, S.; Meiller, T.F. Salivary biomarker profiles in E-cigarette users and conventional smokers: A cross-sectional study. *Oral Dis.* **2021**, *27*, 277–279. [CrossRef] [PubMed]

71. Jacob, P.; St Helen, G.; Yu, L.; Nardone, N.; Havel, C.; Cheung, P.; Benowitz, N.L. Biomarkers of Exposure for Dual Use of Electronic Cigarettes and Combustible Cigarettes: Nicotelline, NNAL, and Total Nicotine Equivalents. *Nicotine Tob. Res. Off. J. Soc. Res. Nicotine Tob.* **2019**, *22*, 1107–1113. [CrossRef] [PubMed]

72. Rudasingwa, G.; Kim, Y.; Lee, C.; Lee, J.; Kim, S.; Kim, S. Comparison of Nicotine Dependence and Biomarker Levels among Traditional Cigarette, Heat-Not-Burn Cigarette, and Liquid E-Cigarette Users: Results from the Think Study. *Int. J. Environ. Res. Public Health* **2021**, *18*, 4777. [CrossRef] [PubMed]

73. Shahab, L.; Goniewicz, M.L.; Blount, B.C.; Brown, J.; McNeill, A.; Alwis, K.U.; Feng, J.; Wang, L.; West, R. Nicotine, Carcinogen, and Toxin Exposure in Long-Term E-Cigarette and Nicotine Replacement Therapy Users: A Cross-sectional Study. *Ann. Intern. Med.* **2017**, *166*, 390–400. [CrossRef] [PubMed]

separations

Article

A Sensitive LC–MS/MS Method for the Quantification of 3-Hydroxybenzo[*a*]pyrene in Urine-Exposure Assessment in Smokers and Users of Potentially Reduced-Risk Products

Nadine Rögner, Heinz-Werner Hagedorn, Gerhard Scherer, Max Scherer and Nikola Pluym *

Analytisch-Biologisches Forschungslabor GmbH, Semmelweisstr. 5, 82152 Planegg, Germany; nadine.roegner@abf-lab.com (N.R.); info@abf-lab.com (H.-W.H.); gerhard.scherer@abf-lab.com (G.S.); max.scherer@abf-lab.com (M.S.)
* Correspondence: pluym@abf-lab.com; Tel.: +49-89-4114796-12

Abstract: Benzo[*a*]pyrene (BaP), a human carcinogen, is formed during the incomplete combustion of organic matter such as tobacco. A suitable biomarker of exposure is the monohydroxylated metabolite 3-hydroxybenzo[*a*]pyrene (3-OH-BaP). We developed a sensitive LC–MS/MS (liquid chromatography coupled with tandem mass spectrometry) method for the quantification of urinary 3-OH-BaP. The method was validated according to the US Food and Drug Administration (FDA) guideline for bioanalytical method validation and showed excellent results in terms of accuracy, precision, and sensitivity (lower limit of quantification (LLOQ): 50 pg/L). The method was applied to urine samples derived from a controlled clinical study to compare exposure from cigarette smoking to the use of potentially reduced-risk products. Urinary 3-OH-BaP concentrations were significantly higher in smokers of conventional cigarettes (149 pg/24 h) compared to users of potentially reduced-risk products as well as non-users (99% < LLOQ in these groups). In conclusion, 3-OH-BaP is a suitable biomarker to assess the exposure to BaP in non-occupationally exposed populations and to distinguish not only cigarette smokers from non-smokers but also from users of potentially reduced-risk products.

Keywords: 3-hydroxybenzo[*a*]pyrene; LC–MS/MS; urine; human biomonitoring; derivatization; potentially reduced-risk products

Citation: Rögner, N.; Hagedorn, H.-W.; Scherer, G.; Scherer, M.; Pluym, N. A Sensitive LC–MS/MS Method for the Quantification of 3-Hydroxybenzo[*a*]pyrene in Urine-Exposure Assessment in Smokers and Users of Potentially Reduced-Risk Products. *Separations* **2021**, *8*, 171. https://doi.org/10.3390/separations8100171

Academic Editor: Fadi Aldeek

Received: 8 September 2021
Accepted: 29 September 2021
Published: 5 October 2021

Publisher's Note: MDPI stays neutral with regard to jurisdictional claims in published maps and institutional affiliations.

1. Introduction

Polycyclic aromatic hydrocarbons (PAHs) are formed during the incomplete combustion of organic matter. High exposures are observed at special workplaces such as cookeries, steel factories, and road buildings. Exposure of the general population to PAHs is mainly caused by environmental factors such as polluted air and water, by the consumption of smoked and grilled food, and by smoking of conventional (combustible) cigarettes (CC), respectively [1–6].

Over the past decade, several new nicotine and tobacco products have been introduced as alternatives to smoking with a potentially reduced health risk compared to CC. As many PAHs are carcinogenic due to their metabolic activation of DNA reactive compounds, the measurement of specific biomarkers is of great importance to assess the exposure to PAHs from potentially reduced-risk products.

For the determination of PAH exposure, usually, respective monohydroxylated urinary metabolites are analyzed by means of LC–MS/MS (liquid chromatography coupled with tandem mass spectrometry) or GC–MS (gas chromatography–mass spectrometry). For instance, 1-hydroxypyrene, monohydroxy-fluorenes, and monohydroxy-phenanthrenes are frequently determined in urine samples in order to investigate exposure to PAHs [7–10]. Benzo[*a*]pyrene (BaP, Figure 1) is classified as a Group 1 carcinogen (carcinogenic to humans) by the International Agency for Research on Cancer (IARC) and is by now the best-studied PAH [3,11].

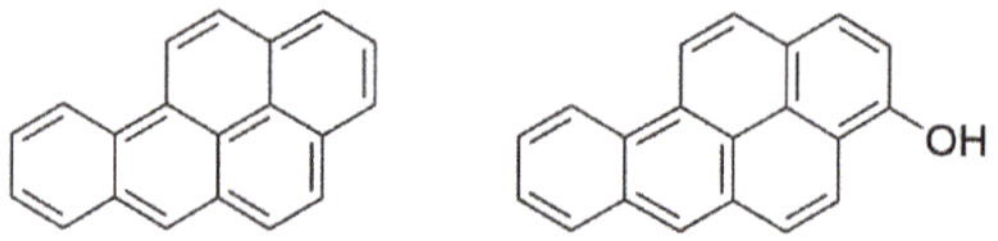

Figure 1. Chemical structure of benzo[*a*]pyrene (BaP, **left**) and its metabolite 3-hydroxy-benzo[*a*]pyrene (3-OH-BaP, **right**).

A key metabolite of BaP is (+)-anti-BaP-7,8-diol-9,10-epoxide (BPDE), which is considered as an ultimate carcinogen, reacting with cellular DNA, proteins, and glutathione. Furthermore, BPDE can react by enzymatic hydrolysis to form BaP-(7,8,9,10)-tetrol, which is excreted in the urine after conjugation with, e.g., glucuronic acid [11–15]. This biomarker, therefore, found use in studying exposure to BaP [16,17]. Very low concentration levels require laborious sample preparation to achieve sufficient sensitivity of the analytical methods, making routine analysis very challenging for this biomarker.

An alternative biomarker of BaP exposure is the monohydroxylated metabolite 3-hydroxybenzo[*a*]pyrene (3-OH-BaP, Figure 1), which is excreted in urine after conjugation. Nearly 100% of the urinary 3-OH-BaP detected in humans is excreted as glucuronide or sulfate [18].

Several methods have been developed and established for the determination of 3-OH-BaP in urine for occupationally exposed subjects [19–23]. However, those methods are limited by the lack of sensitivity to determine the burden of BaP exposure in the general population. In order to cover not only occupational but also environmental exposure, including cigarette smoking, sensitivity in the pg/L-range is required. This can be achieved by optimizing the sample preparation, including derivatization steps [24–27], purification and concentration procedures [28–30], or by application of different ionization techniques, such as atmospheric pressure laser ionization (APLI) [31]. Thus, many of these methods are hampered by complex analytical procedures and specific/expensive equipment that can only be used to a limited extent in larger cohorts of human biomonitoring campaigns and clinical studies, respectively.

The aims of the current study were to adjust and validate a sensitive and robust method for the quantitation of 3-OH-BaP in urine with a sufficiently high sample throughput. Further, the validated method was applied to urine samples collected in a controlled clinical trial [32] with 10 users per group of 5 different nicotine-containing products, including smokers of conventional cigarettes (CC), users of electronic cigarettes (EC), users of heated tobacco products (HTP), users of nicotine replacement therapy (NRT), users of oral tobacco (OT), and non-users (NU), in order to distinguish differences in the exposure from these products.

2. Materials and Methods

2.1. Chemicals

3-Hydroxybenzo[*a*]pyrene-O-β-glucuronide (3-OH-BaP-Gluc, molecular weight (MW): 444 g/mol), 3-OH-BaP (MW: 268 g/mol), and $^{13}C_6$-3-OH-BaP-Gluc were purchased from AptoChem (Montreal, QC, Canada). $^{2}H_{11}$-3-OH-BaP was purchased from TRC (Toronto, ON, Canada). N,N-dimetylethylamine (DMEA) was obtained from Alfa Aesar (Karlsruhe, Germany), 2-fluoro-methylpyridinium-p-toluenesulfonate (FMPT) from TCI (EsMVchborn, Germany), formic acid 99%, ULC/MS grade from Biosolve (Valkenswaard, The Netherlands), and acetic acid, ascorbic acid, dimethyl sulfoxide, hydrochloric acid 37%, and sodium hydroxide from Merck (Darmstadt, Germany). Dichloromethane and methanol for residue analysis and LC–MS grade acetonitrile and methanol were purchased from LGC Standards (Wesel, Germany). Water was purified by means of a Sartorius arium water system (Göttingen, Germany). The enzyme β-glucuronidase/arylsulfatase from *Helix pomatia* (4.5 and 14 U/mL) was supplied by Roche (Mannheim, Germany).

2.2. Sample Work-Up for Quantification

For sample preparation, the work-up published previously [25] was applied with major modifications. Frozen urine samples were thawed slowly at room temperature. To homogenized urine (6 mL), acetate buffer (400 µL; 1 M, pH = 5.1) was added, and the pH-value of the sample was adjusted with hydrochloric acid (1 N) to pH 5.0–5.5 if necessary. Aliquots (100 µL) of an aqueous solution of the internal standard were added, containing an absolute amount of 10 pg $^{13}C_6$-3-OH-BaP-Gluc, followed by the addition of 100 µL of ascorbic acid solution in water (150 mg/mL). For enzymatic hydrolysis, β-glucuronidase/arylsulfatase from *Helix pomatia* (20 µL) was added, and the mixture was incubated overnight (~16–18 h) at 37 °C. After incubation, samples were centrifuged (3000 rpm, 10 min), and the supernatant was decanted into a new vessel and subjected to solid-phase extraction (SPE).

The SPE cartridges (Bond Elut-LMS, 200 mg, 3 mL; Agilent, Waldbronn, Germany) were conditioned with 3 mL of dichloromethane, 2 × 3 mL of methanol, and 3 mL of water. Subsequently, the hydrolyzed urine mixture was added to the column. The tubes were washed with 3 mL of water, 3 mL of water/methanol (50/50, v/v), 1 mL of methanol, and 2 mL of methanol/acetonitrile (50/50, v/v). The target compound and internal standard were eluted with 2 × 2 mL of dichloromethane in a 4 mL glass vial. To the eluate, dimethyl sulfoxide (20 µL) was added, and dichloromethane was evaporated in a SpeedVac centrifuge (Thermo Fisher, Dreieich, Germany) without heating to a final volume of 20 µL (containing only the dimethyl sulfoxide portion).

The residue was taken up in 250 µL of FMPT solution (0.5 mg/mL in acetonitrile) and 50 µL of DMEA (0.2% in acetonitrile). The mixture was homogenized with a vortex mixer, and derivatization of the hydroxyl group was achieved by incubation of the mixture for 20 min at 45 °C. Samples were transferred to a microvial (300 µL), and the solvent was evaporated in a SpeedVac centrifuge (Thermo Fisher, Dreieich, Germany) without heating to a final volume of 20 µL (containing only the dimethyl sulfoxide portion). The residue was reconstituted in 250 µL of methanol/water/formic acid (50/49/1, v/v/v) and homogenized with a vortex mixer. The extracts were analyzed by LC–MS/MS.

2.3. LC–MS/MS

An Agilent 1200 HPLC (Agilent, Waldbronn, Germany) was equipped with an Acquity UPLC BEH C18 column, 50 × 2.1 mm i.d., 1.7 µm (Waters, Eschborn, Germany) and coupled with a triple quadrupole mass spectrometer (API 5000; Sciex, Darmstadt, Germany). The injection volume was set to 15 µL. Chromatography was performed at a column temperature of 50 °C and at a flow rate of 0.6 mL/min. Solvent A (water with 0.5% formic acid) and solvent B (acetonitrile with 0.5% formic acid) were used for elution. The gradient was 0–1 min, 20% B; 1–7 min, 20–40% B; 7–8.5 min, 40% B; 8.5–10 min, 40–90% B; 10–13 min, 90% B; 13–13.1 min, 90–20% B; 13.1–15 min, 20% B. The ion source was operated in electrospray ionization (ESI)-positive mode. Nitrogen was used as the carrier gas. Source parameters were as follows: ion spray voltage, 5500 V; source temperature, 680 °C; entrance potential, 10 V; curtain gas, 30 psi; ion source gas 1, 50 psi; and ion source gas 2, 70 psi. MS measurements were performed by multiple reaction monitoring (MRM) mode. Detailed information for the MRM transitions and MS/MS parameters are summarized in Table 1. For controlling all modules and for data analysis, Analyst 1.5.2 software (Sciex) was used.

Table 1. Retention times, mass transitions, dwell time, declustering potentials (DP), collision energies (CE), and cell exit potentials (CXP) for 3-OH-BaP and $^{13}C_6$-3-OH-BaP.

Analyte or IS	Retention Time (min)	Mass Transitions (*m/z*)	Role	Dwell Time (msec)	DP (V)	CE (V)	CXP (V)
3-OH-BaP	6.9	360 → 251	Quantifier	150	161	45	18
3-OH-BaP	6.9	360 → 267	Qualifier	150	161	45	18
$^{13}C_6$-3-OH-BaP	6.9	366 → 257	IS	150	161	45	18

2.4. Calibration

To determine the concentration of 3-OH-BaP in urine, a calibration line was generated in non-smoker urine (analyte-free) by spiking increasing amounts of 3-OH-BaP-Gluc to receive concentrations between 50 and 3321 pg/L, based on free 3-OH-BaP, while the internal standard amount remained constant (10 pg $^{13}C_6$-3-OH-BaP-Gluc). Calibrators were worked up as described above and analyzed by LC–MS/MS. The calibration line equation was obtained by linear regression ($1/y$ weighting) of the area ratio (area counts of the analyte/area counts of the internal standard) and the spiked analyte concentration. The 3-OH-BaP concentration in human urine samples was then calculated from the area counts ratios between 3-OH-BaP and $^{13}C_6$-3-OH-BaP by employing the calibration line, equation with y being the area count ratio and x being nominal the 3-OH-BaP concentration.

2.5. Method Validation

The method was validated according to the US Food and Drug Administration (FDA) guideline [33]. The method was initially developed and validated using the free forms of both the reference and the internal standard for quantification. Hence, analyte-free non-smoker urine was spiked with free 3-OH-BaP in different concentration levels (low, medium, and high) to cover the entire calibration range. As internal standard, $^2H_{11}$-3-OH-BaP was used during the initial method validation. All working solutions of the analyte and the standard were freshly prepared before use. As a consequence of the stability investigations, the final method comprises the glucuronides 3-OH-BaP-Gluc and $^{13}C_6$-3-OH-BaP instead of the free forms, as discussed in the Section 3. Additionally, ascorbic acid was added to protect the free 3-OH-BaP formed in the urine samples during enzymatic hydrolysis.

To monitor the accuracy and the precision during study sample analysis, internal quality control samples (QCs) were prepared by spiking analyte-free non-smoker urine with known concentrations of 3-OH-BaP-Gluc. The QCs, covering the expected concentration range (QC low, QC medium, QC high), were randomly interspersed with the study samples (min. 5% of total sample size or at least two per level) during sample work-up and analysis. The acceptance criteria for the QCs were defined by accuracy of 85–115%. The target values were previously determined by analyzing six QCs per level.

Selectivity was verified for the MRM transitions of the analyte (quantifier and qualifier) and the corresponding internal standard. Samples of six different analyte-free non-smoker urines were compared with a blank sample containing only the reference compounds, prepared and analyzed under the same conditions. Each transition was screened for potential interferences that had the same retention times as the analyte or the internal standard signal. The same six samples were spiked with 400 pg/L 3-OH-BaP and analyzed for accuracy (85–115%) and precision (CV $\leq$ 15%).

The LLOQ (lower limit of quantification) was determined by analyzing five replicates of spiked non-smoker urine at the lowest concentration (50 pg/L), achieving a precision of at least 20% and an accuracy rate of 80–120%. The LOD (limit of detection) was then obtained by dividing the LLOQ by 3.

Accuracy and precision were determined by spiking non-smoker urine at different concentration levels (LLOQ, low, medium, and high). Inter-day accuracy and precision were determined by analyzing five spiked urine samples per level on three different days. Intra-day accuracy and precision were obtained from the analysis of one day. The acceptance criteria for intra-day and inter-day precision were specified by the calculation of coefficients of variation (CVs), which should be below 15% and 20%, respectively, for concentrations below three times LLOQ. Accuracy rates should be in the range of 85–115% of nominal concentrations and 80–120% for concentrations below three times the LLOQ.

Recovery rates indicate analyte losses during sample work-up. The recovery rates were determined at three different concentration levels by comparing the analyte area of non-smoker urine samples spiked before sample work-up (N = 6) and after SPE extraction (N = 3) with free 3-OH-BaP. Samples spiked after SPE extraction correspond to 100% and served as reference.

The matrix effect (ME) was evaluated by comparing the signals of analyte and internal standard at two different concentration levels (low and high) of post-spiked (after SPE extraction) processed urine samples (N = 3) with a sample of the reference standards. The relative difference to the reference signals (100%) was defined as ME. Relative differences of >0% indicate a positive ME (signal enhancement), and relative differences <0% indicate a negative ME (signal suppression).

Carryover effects were analyzed by repeated injections (N = 3 × 5) of extracts spiked with high levels of the analyte (2000 pg/L) followed by the injection of a blank sample (MeOH). No carryover effects were detected when the signal of the blank sample was at or below the LOD signal.

The stability of the analyte (free 3-OH-BaP or 3-OH-BaP-Gluc) was determined at room temperature for 24 h (short-term stability), at 10 °C for 72 h in the autosampler (post-preparative stability), and below −20 °C (long-term stability). Moreover, six cycles of freeze–thaw stability and the storage stability of stock solutions were monitored. Stability monitoring was performed at two concentration levels (low and medium) in triplicates. Acceptable tolerances were 85–115% compared to the base level (time 0).

2.6. Human Study

The details of the study protocol for the controlled, single-center, and open-label clinical trial has been published previously [32]. All subjects gave their informed consent for inclusion before they participated in the study. The study was conducted in accordance with the Declaration of Helsinki, and the protocol was approved by the Ethics Committee of the Medical Association Hamburg. The study population covered exclusive users of five different nicotine-containing products (CC, EC, HTP, NRT, and OT) and a control group of non-users of any nicotine-containing product (NU). Each group consisted of 10 subjects. Complete urine voids were collected over three days of inpatient stay and pooled to yield 12 h urine samples (12 h periods: from 6 p.m. to 7 a.m. and from 7 a.m. to 6 p.m.). The analysis of 3-OH-BaP comprised the 12 h urine samples on the last day (Day 3, U6 and U7 + 8) of the inpatient stay, as these samples were collected on the third day of confinement under controlled conditions (diet control, habit control), which is the longest time period of control within this study. Main characteristics (user group, sex, age, BMI, and 24 h urine volume of Day 3) of the study population are summarized in Table 2.

For data evaluation, creatinine levels were additionally determined using the Jaffé method [34]. Product use status was verified by the determination of urinary nicotine and its ten metabolites (=total nicotine equivalents, TNE) using SPE (96-well plates) and LC–MS/MS analysis (HILIC column) by modification of a previously published method [35] (Table S1).

2.7. Data Evaluation and Statistics

The statistical parameters were evaluated with Prism (GraphPad, Version 9.0.2, La Jolla, CA, USA). All 3-OH-BaP values below the LLOQ were set to LLOQ/2 (25 pg/L). The urinary 3-OH-BaP concentrations of 12 h urine samples were referred to pg 3-OH-BaP in 24 h (pg/24 h), calculated as concentration 1 × 12 h-urine volume 1 + concentration 2 × 12 h urine volume 2. In addition, analyte concentrations were normalized based on the creatinine concentrations and reported as pg/g creatinine (urinary 3-OH-BaP concentration in pg/L divided by the respective creatinine concentration in g/L). Mean values, standard deviations, and median values were calculated for each user group. Statistical differences between the 3-OH-BaP concentration of smokers and the five other groups were determined by applying the non-parametric Mann–Whitney U test (p-value < 0.05). Statistical differences between the main characteristics of the user groups were determined by applying the non-parametric, one-way ANOVA test (Kruskal–Wallis; p-value < 0.05). Correlation of 3-OH-BaP levels of smokers and smoking-dose-related variables (number of cigarettes smoked per day (CPD) and TNE) were obtained by linear regression and evaluated by calculation of the Spearman correlation coefficient.

Table 2. Main characteristics of the study population.

User Groups [1]	N (m/f)	Age (Years)		BMI		24 h Urine Volume (mL)	
		Mean	±SD	Mean	±SD	Mean	±SD
CC	10 (6/4)	35.1	±9.1	26.0	±3.9	2891	±828
HTP	10 (6/4)	36.1	±12	25.5	±3.2	2685	±1300
OT	10 (9/1)	28.1	±8.2	25.9	±4.2	2638	±1290
EC	10 (6/4)	38.4	±14	23.5	±2.7	1627	±664
NRT	10 (5/5)	35.3	±15	25.5	±3.5	1602	±802
NU	10 (6/4)	32.9	±8.8	24.7	±3.2	2475	±936
Σ all	60 (38/22)	34.3	±11	25.2	±3.4	2320	±1090

[1] User groups: conventional cigarettes (CC), heated tobacco products (HTP), oral tobacco (OT), electronic cigarettes (EC), nicotine replacement therapy (NRT), and non-users (NU).

3. Results

3.1. Performance of the Analytical Method

An LC–MS/MS method published by Sarkar et al. [25] was used as a starting point, further optimized, and finally validated for the quantification of urinary 3-OH-BaP. Sample preparation included enzymatic hydrolysis with glucuronidase/arylsulfatase from *Helix pomatia*, SPE extraction, and derivatization of the hydroxyl group with FMPT (Figure 2). The extracts were then analyzed by LC–MS/MS.

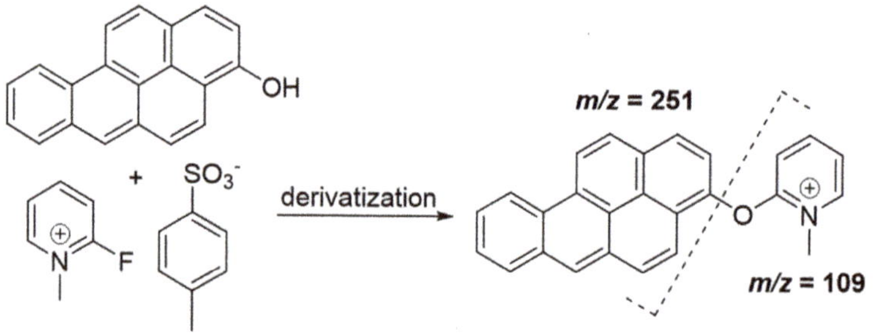

Figure 2. Derivatization of 3-OH-BaP with 2-fluoro-methylpyridinium-p-toluenesulfonate (FMPT).

The final method was validated according to FDA guidelines [33]. The method validation data are shown in Table 3.

Table 3. Method validation data for the quantification of 3-OH-BaP in urine.

Validation Parameter	Level	3-OH-BaP
LOD [1]		16.7 pg/L
LLOQ		50 pg/L
Calibration range		50–3221 pg/L
Precision, intra-day, N = 5		
	LLOQ: 50 pg/L	10.1% CV
	Low: 100 pg/L	12.0% CV
	Medium: 400 pg/L	12.3% CV
	High: 1600 pg/L	3.3% CV
Precision, inter-day, N = 3 × 5		
	LLOQ: 50 pg/L	7.9% CV
	Low: 100 pg/L	9.0% CV
	Medium: 400 pg/L	8.0% CV
	High: 1600 pg/L	5.8% CV
Accuracy, intra-day, N = 5		
	LLOQ: 50 pg/L	101.8%
	Low: 100 pg/L	105.1%
	Medium: 400 pg/L	94.0%
	High: 1600 pg/L	98.2%
Accuracy, inter-day, N = 3 × 5		
	LLOQ: 50 pg/L	105.8%
	Low: 100 pg/L	110.7%
	Medium: 400 pg/L	95.6%
	High: 1600 pg/L	99.6%
Recovery [2,3], N = 6		
	Low: 200 pg/L	121.3%
	Medium: 640 pg/L	108.9%
	High: 1600 pg/L	89.1%
Matrix effect [3], N = 3		
	Low: 200 pg/L	+31.4%
	High: 1600 pg/L	+43.3%
	Low: IS	+25.3%
	High: IS	+47.9%
Re-injection [3], N = 3 × 3		
	Low: 200 pg/L	5.0% CV
	Medium: 640 pg/L	4.6% CV

[1] LOD = LLOQ/3, [2] indicate losses during sample work-up; [3] validation experiments with initial method (cf. 2.5).

The selectivity was proven by analyzing six different analyte-free urine samples. No interfering signals at the same retention times as the analyte or internal standard MRM transitions were detected. Spiking the six samples with the analyte resulted in a mean accuracy of 86.4%.

The precision was evaluated by calculation of the relative standard deviation expressed as CVs, which should not exceed 15% CV (20% CV at LLOQ). Intra-day precision ranged from 3.3% to 12.3% for the different concentration levels. The CVs for the inter-day precision were between 5.8% and 9.0%. The determined intra-day accuracy rates for the LLOQ (101.8%), the low (105.1%), the medium (94.0%), and the high (98.2%) concentration level were within the acceptable range. Inter-day accuracy was also within the range.

For the quantification method, an LLOQ of 50 pg/L was confirmed by the analysis of five independent spiked urine samples on three consecutive days. The LOD was defined as LLOQ/3 and amounted to 16.7 pg/L, showing a signal-to-noise ratio of approximately five. A linear response was found for the calibration range of 50–3221 pg/L.

High recovery rates (89 to 121%) were obtained despite numerous steps in the sample work-up, including enzymatic hydrolysis, SPE extraction, and derivatization. A positive ME in the urine of +31% to +43% was observed for the derivate of 3-OH-BaP. The MEs were fully compensated by the IS. There was no significant carryover, evaluated by a blank sample injected after five consecutive injections of samples with high concentrations. The post-preparative stability of the final extracts was proven in the autosampler at 10 °C for at least 72 h. The reproducibility of re-injection was analyzed by measuring samples with low and medium analyte concentrations in triplicates at three different time points, resulting in CVs of 5.0% and 4.6%, respectively.

Urine samples spiked at two concentration levels with free 3-OH-BaP were stored below −20 °C and analyzed after 1, 3, and 7 days. The accuracy decreased gradually from 106% and 74% on day 1 to 93% and 58% on day 7 for 200 pg/L and 640 pg/L, respectively. Apparently, free 3-OH-BaP was not stable in urine. Degradation was also observed for standard solutions of the analyte and the internal standard in their free form, and thus, fresh solutions needed to be prepared on the day of use. As an alternative for the less stable free 3-OH-BaP, the stability of the glucuronide (3-OH-BaP-Gluc) was investigated as well. 3-OH-BaP-Gluc proved to be stable in urine for at least 30 h at room temperature (short-term stability) and for at least 15 months when stored below −20 °C (long-term stability). The analyte in its conjugated form was stable through six freeze/thaw cycles in urine samples stored below −20 °C. The stock solution of the 3-OH-BaP-Gluc in water (c = 50 µg/mL) was stable for 3.3 years when stored below −20 °C. Consequently, 3-OH-BaP-Gluc was established for the preparation of QC material and for calibration. In analogy, the glucuronide $^{13}C_6$-3-OH-BaP-Gluc was used as an internal standard. Since no interferences were found in the MRM transition, $^{13}C_6$-3-OH-BaP-Gluc was established as IS in the final method to compensate losses during sample work-up. Exemplary MRM chromatograms of non-smoker urine, a QC sample at low concentration, and a smoker urine sample are illustrated in Figure 3.

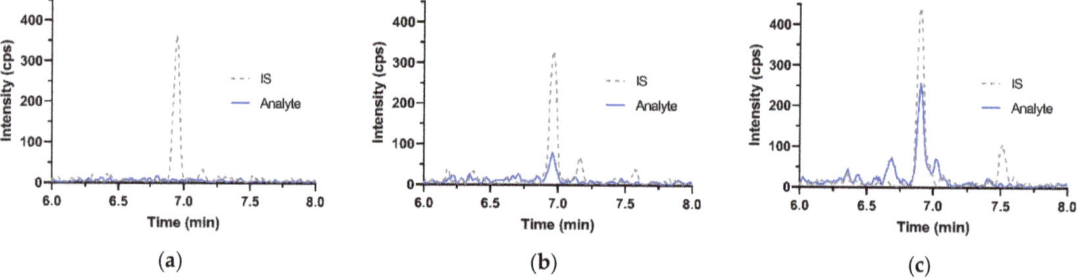

Figure 3. MRM chromatograms of the analyte 3-OH-BaP (m/z 360 → 251) and the internal standard $^{13}C_6$-3-OH-BaP (m/z 366 → 257). (**a**): Non-smoker urine sample (<LOD); (**b**): quality control sample with low concentration (c = 162 pg/L); (**c**): smoker urine sample (c = 470 pg/L).

3.2. Human Study—Urinary Excretion of 3-OH-BaP

The validated method was applied to urine samples from a controlled clinical study [32]. Each group consisted of 10 subjects, resulting in a total number of 60 subjects stratified by product use. The study population was assigned to one of the five groups of users of different nicotine-containing products (CC, EC, HTP, NRT, and OT) based on their product use or to the control group of NU. The main characteristics of the subjects are summarized in Table 2. The confined and diet-controlled clinical study was chosen to ascertain similar (low) exposure to BaP from sources other than product use such as diet or ambient air. Therefore, 3-OH-BaP was quantitated in the 12 h urine samples of the last study day, as this was the longest time period under controlled conditions. Group comparisons were performed based on the total amount of urinary 3-OH-BaP excreted over 24 h (pg/24 h), as summarized in Figure 4 and Table 4.

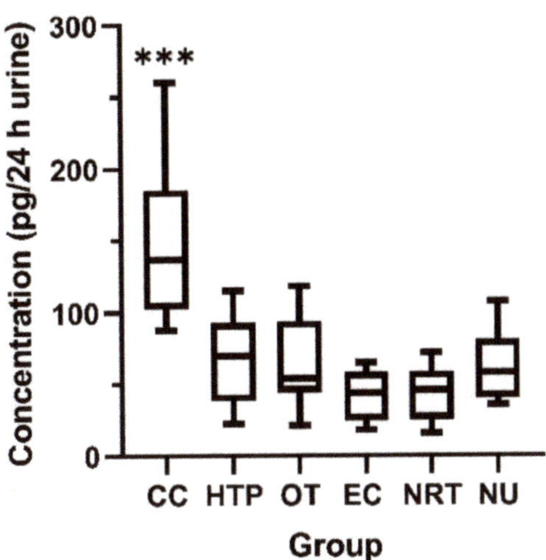

Figure 4. Box plots for urinary 3-OH-BaP excretion (pg/24 h) of six different user groups on Day 3. Boxes and lines represent the range of twenty-fifth/seventy-fifth percentile and the median value. The whiskers illustrate the minimum and maximum concentration. Differences between the smoking (CC) and all other groups were found to be significant when using the non-parametric Mann–Whitney *U* test (***: *p*-value < 0.002).

Table 4. Descriptive statistics for urinary 3-OH-BaP excretion (pg/24 h) of six different user groups.

	3-OH-BaP (pg/24 h) [1]					
	CC	**HTP**	**OT**	**EC**	**NRT**	**NU**
Mean ± SD	149.0 ± 57.0	67.14 ± 32.6	65.96 ± 32.3	40.68 ± 16.6	43.31 ± 19.2	61.88 ± 23.4
Median	136.9	69.40	53.75	43.35	45.40	57.60
Min–max	87.70–260.3	22.30–115.1	21.10–118.0	17.90–64.60	15.30–71.50	35.40–107.3
<LLOQ, N (%) [2]	8 (40%)	20 (100%)	20 (100%)	20 (100%)	19 (95%)	20 (100%)

[1] Levels of 3-OH-BaP excreted within 24 h (N = 10 per group). Concentrations <LLOQ were set to LLOQ/2 (25 pg/L) for calculation of 12 h and 24 h excretion; [2] referred to the concentration in pg/L of 12 h urine samples (N = 20 per group).

The highest mean concentration of 149 pg/24 h was determined for smokers, with 60% of samples above the LLOQ in this group. In contrast, all samples from the other groups, including the NU, were not quantifiable, except for one 12 h urine sample in the NRT group. Mean values varied between groups due to differences in 12 h urine volumes (cf. Table 1). Urinary 3-OH-BaP excretion was significantly higher (*p*-value < 0.002) in smokers compared to all other groups (Figure 4).

3.3. Correlation of 3-OH-BaP with Smoking Specific Parameters

To investigate the specificity of 3-OH-BaP as biomarker of tobacco smoke exposure, urinary 3-OH-BaP levels were plotted against the smoking dose, as indicated by the number of cigarettes smoked per day (CPD) (Figure 5a) and the total nicotine equivalents (TNE) excreted in urine (Figure 5b). Linear regression showed a moderate correlation (of borderline significance) between the 3-OH-BaP concentrations and CPD (Spearman's *r* = 0.63) and a weak (statistically not significant) correlation between 3-OH-BaP concentrations and urinary TNE (Spearman's *r* = 0.52). A reason for the only moderate or weak correlation could be the relatively small sample number. Nevertheless, the positive correlation of 3-OH-BaP with these smoking dose parameters indicates that urinary 3-OH-BaP is a suitable biomarker to assess BaP exposure by cigarette smoking.

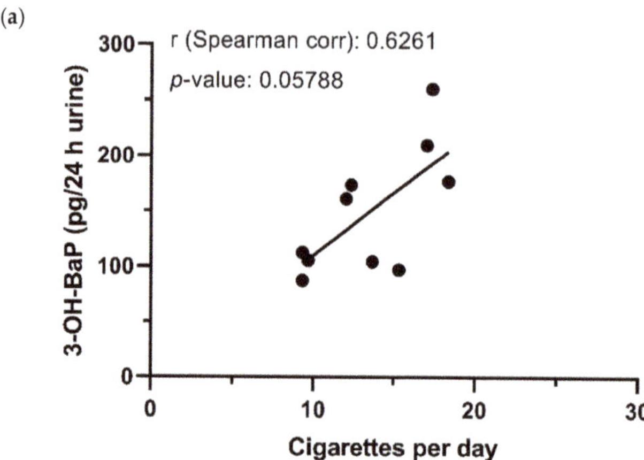

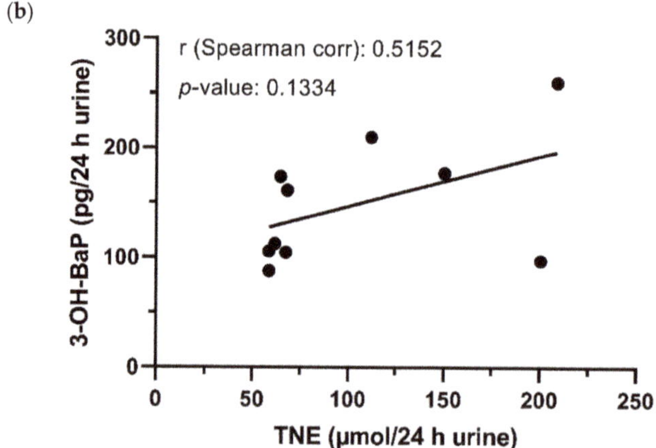

Figure 5. Spearman's correlation between urinary 3-OH-BaP concentrations and (**a**) number of cigarettes smoked per day (CPD) and (**b**) urinary total nicotine equivalents (TNE) of CC users on Day 3 (N = 10).

4. Discussion

4.1. Analytical Method

BaP exposure is most frequently investigated by analyzing the urinary metabolite 3-OH-BaP. Numerous methods have been reported for the quantification of this biomarker in occupationally exposed workers [19–22]. However, these methods generally lack sensitivity for the quantification of 3-OH-BaP in the non-occupationally exposed population. The purpose of the current work was to develop and validate a sensitive method for the quantification of trace amounts of urinary 3-OH-BaP in cohorts of non-occupationally exposed subjects, i.e., in clinical and epidemiological studies. The procedure described by Sarkar et al. [25] was used as a starting point and further modified with respect to the extraction procedure and use of conjugated standards in order to achieve the required sensitivity along with a sufficient sample throughput and robustness. One important improvement in our method in terms of repeatability and accuracy was the implementation of $^{13}C_6$-3-OH-BaP-Gluc as an internal standard. Using the glucuronides as standard and IS material proved to be superior to the unstable free analyte during method validation, calibration, and quality control procedures. In native urine samples, 3-OH-BaP was found to be present in its conjugated form in urine at almost 100% [18]; thus, instability of the analyte would not present an issue in real samples. Another advantage would be that $^{13}C_6$-3-OH-BaP-Gluc could also compensate for losses during enzymatic hydrolysis.

Thus far, only a few methods have been described in the literature that are capable of quantifying BaP exposures in the low pg/L range besides Sarkar et al. [25] and our method. One other method has been published with a similar LLOQ of 50 pg/L by application of liquid chromatography–fluorescence detection (FD) and automated off-line solid-phase extraction [28,36]. While fluorescence detection achieved comparable sensitivity, our method is more specific and selective, applying MRM detection of several analyte-specific mass transitions. Further, LC–MS/MS methods were developed for the quantification of 3-OH-BaP, yielding higher LLOQs. Simon et al. published an automated column-switching high-performance liquid chromatography method for the determination of 3-OH-BaP in urine, yielding quantification limits of approximately 400 pg/L [22,29,30,37]. Several groups analyzed 3-OH-BaP by employing derivatization of the hydroxyl group with dansyl chloride and subsequent analysis by LC–MS/MS, resulting in LLOQs of 250 pg/L [26], 300 pg/L [24], and 580 pg/L [27]. Richter-Brockmann et al. followed a different approach by means of GC–MS using atmospheric pressure laser ionization (APLI). A higher sensitivity compared to our method was reported by the use of APLI, which apparently improved the ionization yields for the methyl ether of 3-OH-BaP with an LLOQ of 1.8 pg/L [31]. Additionally, ascorbic acid was added as an antioxidant before enzymatic cleavage to protect the resulting free 3-OH-BaP from oxidative decomposition during the following work-up and analysis—a procedure that has been established for other PAHs before [38]. Hence, we investigated the addition of ascorbic acid in our method as well. QC samples were analyzed with and without the addition of ascorbic acid, showing no differences in terms of accuracy, sensitivity, and specificity (data not shown). Since the overall variability appeared to be slightly improved by the addition of ascorbic acid, this procedure was implemented into the final method for validation of accuracy, precision, calibration range, and LLOQ and for analysis of the clinical study samples.

4.2. Human Study

The validated method was applied to urine samples of NUs and users of five different nicotine-containing products (CC, EC, NRT, HTP, and OT) who participated in a controlled clinical trial [32]. The aim was not only to investigate whether cigarette smokers and non-smokers differ but also whether cigarette smokers differ from other users of potentially reduced-risk products in terms of exposure to various toxicants, among them BaP. No significant differences were found in terms of the general study group characteristics for age, BMI, and urine volume between the different groups.

The group of CC smokers and NU differ significantly in terms of urinary excretion of 3-OH-BaP. With mean values of 149.0 pg/24 h urine (225.9 pg/g creatinine, Table S2) and 61.88 pg/24 h urine (90.73 pg/g creatinine, Table S2) in CC smokers and NUs, respectively, the determined values were in the same range as reported in the literature [25,28,30].

Barbeau et al. [28] analyzed urinary 3-OH-BaP in non-occupationally exposed non-smokers and smokers. They found an average concentration of 0.009 nmol/mol creatinine for non-smokers and 0.023 nmol/mol creatinine for smokers. These concentrations equaled 45 pg/24 h urine and 155 pg/24 h urine, respectively, assuming a mean 24 h urine volume of 1.5 L and a mean urinary creatinine concentration of 1.5 g/L [28]. The mean concentrations for non-smokers and smokers of 59 pg/24 h urine and 131 pg/24 h urine determined by Lafontaine et al. [30] were in the same range. Sarkar et al. [25] showed, through the analysis of various smoking-specific biomarkers, including 3-OH-BaP, a significant reduction in biomarker concentrations in the group that stopped consuming conventional cigarettes after the baseline of the study. The values at post-baseline were 155 pg/24 h urine for the continuous smoking group and 56 pg/24 h urine for the group that had completely quit using any tobacco product for eight days, which was comparable to non-smokers. In other recent studies analyzing 3-OH-BaP in the urine of non-smokers and smokers, higher concentrations were found in both groups. Concentrations in non-occupationally subjects (N = 4–7) were found in a range of <LOD to 820 pg/L for non-smokers and of 320 to 2150 pg/L for smokers [26,27,31]. Richter-Brockmann et al. speculated that the addition of ascorbic acid may have led to higher concentrations due to improved stability of the analyte. However, our studies using QCs could not prove this hypothesis since comparable concentrations of 3-OH-BaP were observed for identical samples worked-up with and without the addition of ascorbic acid.

In contrast to most studies that investigated exposure in smokers and non-smokers only, Sarkar et al. studied 3-OH-BaP in subjects switching from CC to a snus product. They observed a reduction of about 45% in smokers eight days after switching to snus use, which was in the range of the cessation arm in this study, with a decrease of about 56% [25]. To the best of our knowledge, our clinical study was the first to analyze urinary 3-OH-BaP concentrations to distinguish between smokers and non-smokers, and additionally included four other groups of users of new generation nicotine/tobacco products, such as HTP, OT, EC, and NRT. The four other nicotine user groups could be clearly distinguished from the smokers in terms of their urinary 3-OH-BaP levels, which were indistinguishable from those of NU (99% < LLOQ). The somewhat lower concentrations of EC and NRT users can be explained by the normalization with the 24 h urine volume. It is important to emphasize that the urine volumes collected did not differ significantly between the different user groups (Table 2).

In addition to the number of CPD, urinary cotinine concentration or TNE are commonly used as a biomarker of exposure to nicotine products and have been used for the classification of smoking status. Richter-Brockmann et al. showed a positive correlation of urinary 3-OH-BaP concentration to CPD and cotinine, respectively (R^2 = 0.88 each) [31]. We could confirm the positive correlation between urinary 3-OH-BaP and smoking dose, measured as CPD or TNE in our study (Figure 5).

This study was performed under confined and diet-controlled conditions to reduce the influence of other sources for BaP exposure and to also ascertain compliance of single product use during the inpatient stay. In the clinical study, urine voids were collected over three days, which is regarded as a sufficient time period for the washout of 3-OH-BaP. This was evident, for example, when looking at the progression from Day 1 to Day 3 of a non-compliant NRT user (Figure 6), as identified by the observed CEMA (*N*-acetyl-S-(2-cyanoethyl)-L-cysteine) concentration (a biomarker of exposure to acrylonitrile), although reporting exclusive NRT use for the last three months. Smoking was identified as a major source of acrylonitrile exposure in several studies [39–41]. A CEMA cut-off between 0.4 and 0.7 µg/L was recently suggested [42]. The non-compliant subject in the NRT user group showed a CEMA concentration of 86 µg/L on Day 1 (U0) equal to 84 µg/12 h urine fraction, strongly indicating cigarette smoking before the study started.

The high 3-OH-BaP concentration (>200 pg/12 h urine) in the U0 fraction supports these findings. A downward trend was observed from Day 1 to Day 3, indicating that non-compliant behavior and other sources of BaP exposure could be excluded during the course of the study. Despite possible non-compliance and other BaP exposure sources, NUs and all other nicotine user groups were significantly distinguishable from CC smokers in the urine samples collected before the study started (U0) (*p*-value < 0.05, Table S3). These findings show that 3-OH-BaP is significantly elevated in smokers in an uncontrolled setting as well, emphasizing the suitability of 3-OH-BaP as a biomarker to discriminate cigarette smoking from other nicotine-containing products such as e-cigarettes, smokeless and oral tobacco, or heated tobacco products. Since only a small sample size of each user group (N = 10) was used here, these findings would need to be confirmed in larger cohorts.

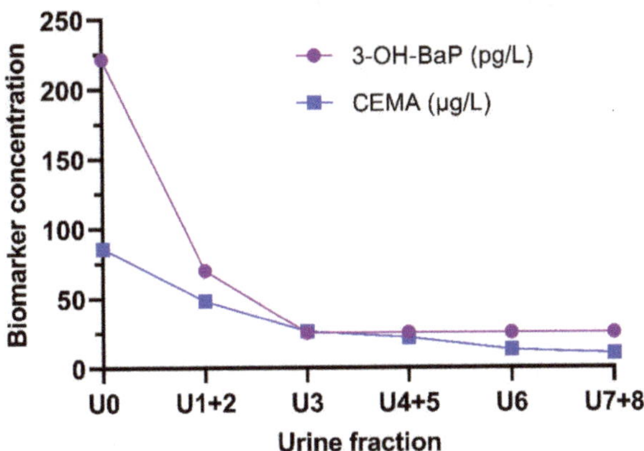

Figure 6. Progression of 3-OH-BaP (pg/L) and CEMA (µg/L) concentration of an NRT user over the three study days.

5. Conclusions

The new LC–MS/MS method is highly sensitive and allows for quantification of urinary 3-OH-BaP in cohorts of non-occupationally exposed subjects due to high throughput. Covering a broad, linear calibration range and an LLOQ of 50 pg/L, the actual method is suitable for the quantification of occupationally and non-occupationally exposed populations. Smokers can be differentiated from non-smokers as well as from users of new generation tobacco/nicotine and oral tobacco products. A moderate correlation between urinary 3-OH-BaP and the smoking dose was observed. Hence, 3-OH-BaP is a suitable biomarker to discriminate smokers from users of potentially reduced-risk products. The method is also suitable for assessing low exposures to BaP originating from diet and ambient air.

Supplementary Materials: The following are available online at https://www.mdpi.com/article/10.3390/separations8100171/s1, Table S1: Descriptive statistics for TNE (µmol/24 h urine) of six different user groups on Day 3, Table S2: Descriptive statistics for urinary 3-OH-BaP excretion (pg/g creatinine) of six different user groups on Day 3, Table S3: Descriptive statistics for urinary 3-OH-BaP excretion (pg/12 h urine) of six different user groups on Day-1.

Author Contributions: Conceptualization, M.S., G.S. and N.P.; methodology, N.R. and H.-W.H.; validation, N.R. and H.-W.H.; formal analysis, N.R.; investigation, N.R.; writing—original draft preparation, N.R.; writing—review and editing, M.S., G.S. and N.P.; supervision, N.P.; project administration, N.P.; funding acquisition, M.S. All authors have read and agreed to the published version of the manuscript.

Funding: This study was funded with a grant from the Foundation for a Smoke-Free World, Inc. ("FSFW"), a US nonprofit 501 (c) (3) private foundation. This study is, under the terms of the grant agreement with FSFW, editorially independent of FSFW. The contents, selection, and presentation of facts, as well as any opinions expressed herein, are the sole responsibility of the authors and under no circumstances shall be regarded as reflecting the positions of the FSFW. The FSFW accepts charitable gifts from PMI Global Services Inc. (PMI); under FSFW's Bylaws and Pledge Agreement with PMI, the FSFW is independent of PMI and the tobacco industry.

Institutional Review Board Statement: The study was conducted according to the guidelines of the Declaration of Helsinki, and the protocol was approved by the Ethics Committee of the Medical Association Hamburg (reference number: PV7084, date of approval: 10 September 2019).

Informed Consent Statement: Informed consent was obtained from all subjects involved in the study.

Data Availability Statement: The data presented here are available on request from the corresponding author.

Acknowledgments: The authors thank CTC North GmbH (Hamburg, Germany) for conducting the clinical study, and our colleagues from ABF for their assistance in organizing and managing the clinical study, and for analysis of the urinary nicotine metabolites.

Conflicts of Interest: The authors declare no conflict of interest. The funders had no role in the design of the study; in the collection, analyses, or interpretation of data; in the writing of the manuscript, or in the decision to publish the results.

References

1. Lijinsky, W. The formation and occurrence of polynuclear aromatic hydrocarbons associated with food. *Mutat. Res. Genet. Toxicol.* **1991**, *259*, 251–261. [CrossRef]
2. Hattemer-Frey, H.A.; Travis, C.C. Benzo-a-pyrene: Environmental partitioning and human exposure. *Toxicol. Ind. Health* **1991**, *7*, 141–157. [CrossRef] [PubMed]
3. IARC. *Monographs on the Evaluation of Carcinogenic Risks to Humans. Some Non-Heterocyclic Polycyclic Aromatic Hydrocarbons and Some Related Exposures*; IRAC: Lyon, France, 2010; Volume 92.
4. Rodgman, A.; Perfetti, T. The Composition of Cigarette Smoke: A Catalogue of the Polycyclic Aromatic Hydrocarbons. *Contrib. Tob. Res.* **2006**, *22*, 13–69. [CrossRef]
5. IARC. *Monographs on the Evaluation of Carcinogenic Risks to Humans. Polynuclear Aromatic Compounds, Part 1: Chemical, Environmental and Experimental Data*; IRAC: Lyon, France, 1983; Volume 32.
6. IARC. *Monographs on the Evaluation of Carcinogenic Risks to Humans. Chemical Agents and Related Occupations*; IRAC: Lyon, France, 2012; Volume 100 F.
7. Scherer, G.; Frank, S.; Riedel, K.; Meger-Kossien, I.; Renner, T. Biomonitoring of Exposure to Polycyclic Aromatic Hydrocarbons of Nonoccupationally Exposed Persons. *Cancer Epidemiol. Biomark. Prev.* **2000**, *9*, 373–380.
8. Jacob, J.; Seidel, A. Biomonitoring of polycyclic aromatic hydrocarbons in human urine. *J. Chromatogr. B: Anal. Technol. Biomed. Life Sci.* **2002**, *778*, 31–47. [CrossRef]
9. Ramsauer, B.; Sterz, K.; Hagedorn, H.W.; Engl, J.; Scherer, G.; McEwan, M.; Errington, G.; Shepperd, J.; Cheung, F. A liquid chromatography/tandem mass spectrometry (LC-MS/MS) method for the determination of phenolic polycyclic aromatic hydrocarbons (OH-PAH) in urine of non-smokers and smokers. *Anal. Bioanal. Chem.* **2011**, *399*, 877–889. [CrossRef] [PubMed]
10. Dor, F.; Dab, W.; Empereur-Bissonnet, P.; Zmirou, D. Validity of biomarkers in environmental health studies: The case of PAHs and benzene. *Crit. Rev. Toxicol.* **1999**, *29*, 129–168. [CrossRef]
11. Deutsche Forschungsgemeinschaft (DFG). Benzo[a]pyrene. In *MAK Collection: Occupational toxicants, Part 1*; Commission, M., Ed.; Wiley-VCH Verlag: Heidelberg, Germany, 2012; Volume 27.
12. Conney, A.H.; Chang, R.L.; Jerina, D.M.; Caroline Wei, S.J. Studies on the Metabolism of Benzo[a]Pyrene and Dose-Dependent Differences in the Mutagenic Profile of Its Ultimate Carcinogenic Metabolite. *Drug Metab. Rev.* **1994**, *26*, 125–163. [CrossRef]
13. Gelboin, H.V. Benzo[a]pyrene metabolism, activation and carcinogenesis: Role and regulation of mixed-function oxidases and related enzymes. *Physiol. Rev.* **1980**, *60*, 1107–1166. [CrossRef]
14. Andreas, L.; William, M.B. Metabolic Activation and Detoxification of Polycyclic Aromatic Hydrocarbons. In *The Carcinogenic Effects of Polycyclic Aromatic Hydrocarbons*; World Scientific: London, UK, 2005; pp. 19–96.
15. Verma, N.; Pink, M.; Rettenmeier, A.W.; Schmitz-Spanke, S. Review on proteomic analyses of benzo[a]pyrene toxicity. *Proteomics* **2012**, *12*, 1731–1755. [CrossRef]
16. Zhong, Y.; Carmella, S.G.; Hochalter, J.B.; Balbo, S.; Hecht, S.S. Analysis of r-7,t-8,9,c-10-Tetrahydroxy-7,8,9,10-tetrahydrobenzo[a]pyrene in Human Urine: A Biomarker for Directly Assessing Carcinogenic Polycyclic Aromatic Hydrocarbon Exposure Plus Metabolic Activation. *Chem. Res. Toxicol.* **2011**, *24*, 73–80. [CrossRef] [PubMed]
17. Richter-Brockmann, S.; Dettbarn, G.; Jessel, S.; John, A.; Seidel, A.; Achten, C. GC-APLI-MS as a powerful tool for the analysis of BaP-tetraol in human urine. *J. Chromatogr. B* **2018**, *1100–1101*, 1–5. [CrossRef] [PubMed]
18. Luo, K.; Gao, Q.; Hu, J. Determination of 3-Hydroxybenzo[a]pyrene Glucuronide/Sulfate Conjugates in Human Urine and Their Association with 8-Hydroxydeoxyguanosine. *Chem. Res. Toxicol.* **2019**, *32*, 1367–1373. [CrossRef]
19. Gündel, J.; Angerer, J. High-performance liquid chromatographic method with fluorescence detection for the determination of 3-hydroxybenzo[a]pyrene and 3-hydroxybenz[a]anthracene in the urine of polycyclic aromatic hydrocarbon-exposed workers. *J. Chromatogr. B Biomed. Sci. Appl.* **2000**, *738*, 47–55. [CrossRef]
20. Gündel, J.; Schaller, K.H.; Angerer, J. Occupational exposure to polycyclic aromatic hydrocarbons in a fireproof stone producing plant: Biological monitoring of 1-hydroxypyrene, 1-, 2-, 3- and 4-hydroxyphenanthrene, 3-hydroxybenz(a)anthracene and 3-hydroxybenzo(a)pyrene. *Int. Arch. Occup. Environ. Health* **2000**, *73*, 270–274. [CrossRef]
21. Raponi, F.; Bauleo, L.; Ancona, C.; Forastiere, F.; Paci, E.; Pigini, D.; Tranfo, G. Quantification of 1-hydroxypyrene, 1- and 2-hydroxynaphthalene, 3-hydroxybenzo[a]pyrene and 6-hydroxynitropyrene by HPLC-MS/MS in human urine as exposure biomarkers for environmental and occupational surveys. *Biomarkers* **2017**, *22*, 575–583. [CrossRef]

22. Förster, K.; Preuss, R.; Rossbach, B.; Bruning, T.; Angerer, J.; Simon, P. 3-Hydroxybenzo[a]pyrene in the urine of workers with occupational exposure to polycyclic aromatic hydrocarbons in different industries. *Occup. Environ. Med.* **2008**, *65*, 224–229. [CrossRef]

23. Gendre, C.; Lafontaine, M.; Morele, Y.; Payan, J.-P.; Simon, P. Relationship Between Urinary Levels of 1-Hydroxypyrene and 3-Hydroxybenzo[a]pyrene for Workers Exposed to Polycyclic Aromatic Hydrocarbons. *Polycyclic Aromat. Compd.* **2002**, *22*, 761–769. [CrossRef]

24. Luo, K.; Gao, Q.; Hu, J. Derivatization method for sensitive determination of 3-hydroxybenzo [a] pyrene in human urine by liquid chromatography–electrospray tandem mass spectrometry. *J. Chromatogr. A* **2015**, *1379*, 51–55. [CrossRef]

25. Sarkar, M.; Liu, J.; Koval, T.; Wang, J.; Feng, S.; Serafin, R.; Jin, Y.; Xie, Y.; Newland, K.; Roethig, H.J. Evaluation of biomarkers of exposure in adult cigarette smokers using Marlboro Snus. *Nicotine Tob. Res.* **2010**, *12*, 105–116. [CrossRef] [PubMed]

26. Yao, L.; Yang, J.; Liu, B.; Zheng, S.; Wang, W.; Zhu, X.; Qian, X. Development of a sensitive method for the quantification of urinary 3-hydroxybenzo[a]pyrene by solid phase extraction, dansyl chloride derivatization and liquid chromatography-tandem mass spectrometry detection. *Anal. Methods* **2014**, *6*, 6488–6493. [CrossRef]

27. Hu, H.; Liu, B.; Yang, J.; Lin, Z.; Gan, W. Sensitive determination of trace urinary 3-hydroxybenzo[a]pyrene using ionic liquids-based dispersive liquid-liquid microextraction followed by chemical derivatization and high performance liquid chromatography-high resolution tandem mass spectrometry. *J. Chromatogr. B* **2016**, *1027*, 200–206. [CrossRef]

28. Barbeau, D.; Maître, A.; Marques, M. Highly sensitive routine method for urinary 3-hydroxybenzo[a]pyrene quantitation using liquid chromatography-fluorescence detection and automated off-line solid phase extraction. *Analyst* **2011**, *136*, 1183–1191. [CrossRef]

29. Simon, P.; Lafontaine, M.; Delsaut, P.; Morele, Y.; Nicot, T. Trace determination of urinary 3-hydroxybenzo[a]pyrene by automated column-switching high-performance liquid chromatotgraphy. *J. Chromatogr. B Biomed. Sci. Appl.* **2000**, *748*, 337–348. [CrossRef]

30. Lafontaine, M.; Champmartin, C.; Simon, P.; Delsaut, P.; Funck-Brentano, C. 3-Hydroxybenzo[a]pyrene in urine of smokers and non-smokers. *Toxicol. Lett.* **2006**, *162*, 181–185. [CrossRef]

31. Richter-Brockmann, S.; Dettbarn, G.; Jessel, S.; John, A.; Seidel, A.; Achten, C. Ultra-high sensitive analysis of 3-hydroxybenzo[a]pyrene in human urine using GC-APLI-MS. *J. Chromatogr. B* **2019**, *1118–1119*, 187–193. [CrossRef]

32. Sibul, F.; Burkhardt, T.; Kachhadia, A.; Pilz, F.; Scherer, G.; Scherer, M.; Pluym, N. Identification of biomarkers specific to five different nicotine product user groups: Study protocol of a controlled clinical trial. *Contemp. Clin. Trials Commun.* **2021**, *22*, 100794. [CrossRef]

33. Food and Drug Administration (FDA). Bioanalytical Method Validation—Guidance for Industry. FDA-2013-D-1020; 2018. Available online: https://www.fda.gov/regulatory-information/search-fda-guidance-documents/bioanalytical-method-validation-guidance-industry (accessed on 6 September 2021).

34. Blaszkewicz, M.; Liesenhoff-Henze, K. Creatinine in urine [Biomonitoring Methods, 2010]. In *The MAK-Collection for Occupational Health and Safety*; Wiley-VCH: Weinheim, Germany, 2010; pp. 169–184.

35. Piller, M.; Gilch, G.; Scherer, G.; Scherer, M. Simple, fast and sensitive LC-MS/MS analysis for the simultaneous quantification of nicotine and 10 of its major metabolites. *J. Chromatogr. B* **2014**, *951–952*, 7–15. [CrossRef]

36. Barbeau, D.; Persoons, R.; Marques, M.; Hervé, C.; Laffitte-Rigaud, G.; Maitre, A. Relevance of urinary 3-hydroxybenzo (a) pyrene and 1-hydroxypyrene to assess exposure to carcinogenic polycyclic aromatic hydrocarbon mixtures in metallurgy workers. *Ann. Occup. Hyg.* **2014**, *58*, 579–590. [CrossRef]

37. Lafontaine, M.; Gendre, C.; Delsaut, P.; Simon, P. Urinary 3-hydroxybenzo[a]pyrene as a biomarker of exposure to polycyclic aromatic hydrocarbons: An approach for determining a biological limit value. *Polycycl. Aromat. Compd.* **2004**, *24*, 441–450. [CrossRef]

38. Li, Z.; Romanoff, L.C.; Trinidad, D.A.; Pittman, E.N.; Hilton, D.; Hubbard, K.; Carmichael, H.; Parker, J.; Calafat, A.M.; Sjödin, A. Quantification of 21 metabolites of methylnaphthalenes and polycyclic aromatic hydrocarbons in human urine. *Anal. Bioanal. Chem.* **2014**, *406*, 3119–3129. [CrossRef] [PubMed]

39. Alwis, K.U.; Blount, B.C.; Britt, A.S.; Patel, D.; Ashley, D.L. Simultaneous analysis of 28 urinary VOC metabolites using ultra high performance liquid chromatography coupled with electrospray ionization tandem mass spectrometry (UPLC-ESI/MSMS). *Anal. Chim. Acta* **2012**, *750*, 152–160. [CrossRef] [PubMed]

40. Pluym, N.; Gilch, G.; Scherer, G.; Scherer, M. Analysis of 18 urinary mercapturic acids by two high-throughput multiplex-LC-MS/MS methods. *Anal. Bioanal. Chem.* **2015**, *407*, 5463–5476. [CrossRef] [PubMed]

41. Schettgen, T.; Musiol, A.; Alt, A.; Ochsmann, E.; Kraus, T. A method for the quantification of biomarkers of exposure to acrylonitrile and 1,3-butadiene in human urine by column-switching liquid chromatography-tandem mass spectrometry. *Anal. Bioanal. Chem.* **2009**, *393*, 969–981. [CrossRef] [PubMed]

42. Luo, X.; Carmella, S.G.; Chen, M.; Jensen, J.A.; Wilkens, L.R.; Le Marchand, L.; Hatsukami, D.K.; Murphy, S.E.; Hecht, S.S. Urinary Cyanoethyl Mercapturic Acid, a Biomarker of the Smoke Toxicant Acrylonitrile, Clearly Distinguishes Smokers From Nonsmokers. *Nicotine Tob. Res.* **2020**, *22*, 1744–1747. [CrossRef] [PubMed]

MDPI
St. Alban-Anlage 66
4052 Basel
Switzerland
Tel. +41 61 683 77 34
Fax +41 61 302 89 18
www.mdpi.com

Separations Editorial Office
E-mail: separations@mdpi.com
www.mdpi.com/journal/separations